Reconstituting India

Reconstituting India

Edited by

PRADIP SARBADHIKARI

Calcutta
Oxford University Press
Delhi Bombay Madras
1995

Oxford University Press, Walton Street, Oxford OX2 6DP

OXFORD NEW YORK
ATHENS AUCKLAND BANGKOK BOMBAY
CALCUTTA CAPE TOWN DAR ES SALAAM DELHI
FLORENCE HONG KONG ISTANBUL KARACHI
KUALA LUMPUR MADRAS MADRID MELBOURNE
MEXICO CITY NAIROBI PARIS SINGAPORE
TAIPEI TOKYO TORONTO

and associates in
BERLIN IBADAN

ISBN 019 5637003

Printed in India
at Asutosh Lithographic Co., 13, Chidam Mudi Lane, Calcutta 700 006
and published by Neil O'Brien, Oxford University Press,
5 Lala Lajpat Rai Sarani, Calcutta 700 020

Respectfully dedicated to the memory of
my father, Dr Prabhat Chandra Sarbadhikari (1898-1953);
my mother, Mrs Sovona Sarbadhikari (1907-1984); and
my sister, Mrs Priti Ghose (1931-1984)
for their infinite love, care and high ideals

About the Contributors

Pradip Sarbadhikari studied at Presidency College, Calcutta, the London School of Economics and Political Science, and the Royal University of Utrecht, The Netherlands. He has taught and lectured widely in Asia, Europe and America. He has over three decades actively promoted the study of India in the West through building of organizations and participation in numerous academic bodies. He is Professor and Chairman, Department of Political Studies in Lakehead University, Ontario, Canada. He was a member of the National Advisory Panel to the Minister of External Relations, Government of Canada for selecting Centres of Excellence in International Development at Canadian universities. He is the author of numerous papers and books including *India and the Great Powers* and *New Nationalism and International Society*.

P M Baxi was a member of the Indian Law Commission and its Honorary Director. He is co-editor of Dr B. K. Mukherjee's Tagore Law Lectures on Hindu Religious Endowments and other books and articles on property, procedure and constitutional law.

Gurcharan Singh Bhalla studied at the London School of Economics and Political Science, and is Professor of Economics at Jawaharlal Nehru University specializing in Agricultural Economics. He is a former Chairman of the Agricultural Prices Review Commission and a member of the Planning Commission, Government of India.

Dhires Bhattacharya taught Economics at Presidency College, Calcutta, and the Universities of Calcutta and Wellington (New Zealand). He was associated with the Research Programmes Committee of the Planning Commission. He has written extensively on Indian economic planning and policies and has published four books in the field.

Biplab Dasgupta is Professor of Economics at Calcutta University and a Member of the West Bengal State Planning Board. He studied at the London School of Economics and Political Science, and taught at London University (School of Oriental and African Studies) and at the University of Sussex. He was Vice Chairman of the West Bengal Comprehensive Area Development Corporation, initiating and overseeing rural development. He also served and is currently a Member of Parliament. He is the author of several papers and books including *The Oil Industry in India*, *The Naxalite Movement*, *Agrarian Change and New Technology in India*, and a pioneering methodological study in co-authorship with Professor Morris Jones: *Patterns and Trends in Indian Politics.*

Arun Ghosh joined the Government of India in 1950 prior to which he taught Economics at the University of Lucknow. He was Economic Advisor to the Government of India, Alternate Executive Director (for India) on the Boards of the World Bank and IMF, and Secretary to the Government of India. Lately, he has been Vice-Chairman, West Bengal State Planning Board, and Member of the Planning Commission, Government of India.

Baharul Islam was a former Chief Justice of Guwahati High Court, Assam, and former Justice of the Supreme Court of India. He was also a Member of Parliament. His decisions and writings have evoked keen debate and have been widely read. The editor deeply regrets Justice Islam's sudden death. His contribution to this volume is one of his last essays.

Rajni Kothari is author, among other works, of *State against* Democracy, Transformation and Survival and *Rethinking* Development, and editor of *State and Nation Building: A Third World Perspective*. A graduate of the London School of Economics and Political Science, he founded the Centre for the Study of Developing Societies, Delhi in 1963 of which he was Director for several years. In 1975 he started the well-known international journal, *Alternatives*. He is a former President of the Peoples' Union of Civil Liberties and a former member of the Planning Commission, Government of India.

Madhu Limaye is a freedom fighter. He held various offices in the Socialist movement, such as Chairman of the Party and Leader of the Parliamentary Group. He was a member of the third, fourth, fifth and sixth Lok Sabhas. Since then he has written several books and articles.

Amit Mitra studied at Duke University, USA, and taught at the Indian Institute of Technology, Delhi, and the Franklin and Marshall College, Pennsylvania, USA. He is Director of the Policy Unit of the *Business India* Group and writes a monthly column for *Business India*. He is also Senior Consultant with the Indian Council for Research on International and Economic Relations at Delhi. He has developed a rural project in remote Bihar and held political positions.

Pranab Mukherjee is a Cabinet Minister holding the portfolio of Finance in the Cabinet of Prime Minister Indira Gandhi and of Commerce, and currently, External Affairs, in the Cabinet of Prime Minister P. V. Narasimha Rao. He has held several leading party positions and is a Member of Parliament, and Deputy Chairman of the Planning Commission.

Satish Saberwal studied at Vanderbilt University, USA, and is Professor of Sociology, Centre for Historical Studies, Jawaharlal Nehru University, New Delhi. He has done anthropological field work in Kenya and in Punjab, and his recent interests have included medieval Europe. His *Wages of Segmentation: Comparative Studies on Europe and India over the Long Term,* is being published by Oxford University Press. He is the author of several other books, monographs and articles including *India in Crisis.*

Contents

Preface

Almost half a century of political sovereignty in India has brought into sharp relief her successes and failures, highlighting the nature of the Indian State and its economic and social reforms. To address the issues implicit in this transformation I asked leading Indian thinkers with diverse ideological, academic and professional backgrounds to provide contending views on changes in India. My introductory chapter 'Transition in India' sketches the main arguments on state, economy, society and the world order as a background to the essays by eleven distinguished scholars, political personalities and policy-makers who deal with the process of transition.

I owe a special debt of gratitude to an old school friend from St Xavier's Calcutta, for sustained support for the project and to my classmate at Presidency College, Calcutta, Dr Amiya Kumar Bagchi for his helpful criticism and encouragement. I am also obliged to several others: to each of the illustrious contributors for taking time off to write for this volume; to Lakehead University, Ontario, Canada for giving me freedom over many years to remain in close contact with changes in India, and for a sabbatical leave which gave me the opportunity to distil some ideas in this book; to my friends and students in the West and in India who over the years contributed to my evolving ideas on India's predicaments; and finally, to my wife Conchita who shares my life completely and who instilled in my daughter Palomita and sons Probir and Prodypto, love for their Indian heritage.

84, Ballygunge Place
Calcutta 700 019, India
1st July 1994

PRADIP SARBADHIKARI

Introduction

PRADIP SARBADHIKARI

Recent changes in India's domestic politics and the international political system together with concurrent moves towards economic globalization have led to the adoption of new economic policies in India which reflect a distinct shift from those pursued since Independence. While it is too early to examine the impact of these changes and determine whether they constitute a socio-economic transformation or not, there is an urgent necessity for serious consideration of not only these policies *per se* but also of the political and social context in which they are being discussed. It is with this objective that the present study has sought to incorporate changes in the State, the economy, and society in India. Despite a plethora of articles on the State, much of the current debate on the new economic policy has directly focused on the merits and shortcomings of liberalization to the exclusion of an examination of the Indian State with reference to changes since Independence in India's political economy. Such analyses are essential for explaining the significance of the new globally conceived economic policies, the new social stirrings of identity, and indicate the path India will take in the twenty-first century. Rajni Kothari focuses upon this central concern in the chapter on 'State in Transition' saying: 'Of all the uncertainties facing India perhaps the most critical is with respect to the condition and the role of the State' and concludes: 'It is important for us to understand the pathology of a development process based on the market before we decide to throw away the advantages of a comprehensive process of nation-building and a corresponding political economy in which a central role is assigned to the State'. This analysis is complemented by Madhu Limaye in 'Indian Polity Today' in which he details the constitutional and political evolution shaping the political, administrative and judicial structure and the nature of the crisis facing the Indian polity. Almost half a century after Partition, the problem of sub-nationalism and various types of self-determination pose the gravest challenge to independent India's unity. Addressing the issue of subnationalism in the context of India's constitutional structure, P.M. Baxi explores the nature of India's federalism and the place of sub-nationalism within it. The ideological make up of the Indian State and of its leaders and their economic impact is the theme of Biplab Dasgupta's study. In 'State Power and Indigenous Capitalist Development' he argues, 'I take the view that the path pursued by Nehru was not a socialist one; on the contrary, his objective was to build capitalism, and it failed because the Indian State was unable to fulfil the necessary conditions for such indigenous capitalist development e.g. implementation of land reform'.

Dhires Bhattacharya provides an overview of the main features of the economic situation since Independence drawing his observations from statistical data. He then introduces two sides of the debate: 'those who see freedom to invest as leading to hegemonic control over the economic system [by] some large industrialist' and, 'those who feel that only through growth can the unemployed and poverty-stricken be drawn into the economic mainstream'.

Focusing on the crucial agricultural sector, Gurcharan Singh Bhalla in 'Agrarian Change in India since Independence' examines the 'changes in the agrarian structure as a result of land reform legislation, the variations in the nature and extent of the spatial pattern, and a discussion of the new contradictions and rural tensions in present-day India, between the increasing mass of landless cultivators and the large cultivators'. The resolution lies in the ability of the democratic movement to bring about the required restructuring of the Indian economy and polity.

The disintegration of the Soviet socialist system and the simultaneous policies of privatization closely linked to the World Bank's call for structural adjustment, has placed the very fabric of planning in India in great jeopardy. Pranab Mukherjee responds to this critical danger in his article 'Planning in a Liberalized Economy' by asserting that, 'those who put planning and the market as alternative systems, perhaps, do not realize that the two could be best seen and made complementary to each other'. Specifically applying his analysis to the Eighth Plan, he warns that the public sector 'may not be expected to maintain its conventional share in investment subsidies which need to be rationalized ... once a macroeconomic framework is determined, planning should be done on a selective sector-specific basis'.

This shift in economic policy is seen as a recipe for subversion of India's economic sovereignty without producing the 'promised bonanza growth in the long run'. Arun Ghosh states this position in 'Perspectives and Recent Policy Changes' and warns that 'the present programme of adjustment and structural reform contains within it elements of self-destruction'. He explains the intensification of unemployment, the curtailment of social security programmes and the danger of the 'conflict between short-term needs of adjustment and medium-term objectives of expanding output and increasing efficiency in production, falling into the debt trap'. 'India is today facing a crisis which has spilled over to the political arena and threatens to tear the country apart.' The nature and implications of industrial liberalization which are the hallmark of the post-Rajiv Gandhi reforms is the subject of Amit Mitra's paper. In an epilogue that updates the key reforms, he sees the State as 'a facilitator of private business activities, rather than that of judge and jury' but warns that 'information processing and a high degree of analytical skills will be demanded of Government functionaries'.

While explanations of the State's transition are linked to economic policies, the evolution of Indian society and the nature of changes and obstacles to change have been addressed by Satish Saberwal in 'Reconstituting India'. In the backdrop of 'the happenings of the former Soviet Union ... we are all caught in a vast historic process .... It is Western Europe that worked out early, the institutions and technologies which today set the terms for everyone else ... Vastly enlarged scales, of possible social relationships, of production, and of much else, are central to this evolving universe. Alternate visions, whether of the Marxists or the Third World nationalists have been losing steam; and as their protagonists try headlong to catch up with the West, they find themselves stumbling too often.' This is the general milieu within which one has to locate the particular Indian case. Of the many crises facing 'this Indian case', one of the gravest is religious fundamentalism. This discussion is complemented by Baharul Islam's examination of the status of Muslims in 'The State, Personal Law and Islam in India'. This study explores in depth the various aspects of personal law as it pertains to marriage, divorce, maintenance, inheritance, diversity of interpretations and its relationship to the Indian State and other religious systems.

These essays reflect the core argument about what the proper relationship is between the State and the market. They offer alternative perspectives on the nature, evolution, ideological bias and structural imperatives of the State; contending judgements on the impact of economic restructuring; responses to critical social concerns; and the changed global environment for policy-making. These deliberations and the implicit predications arising from them would inevitably promote or hinder India's welfare as she enters the twenty-first century.

As the reforms in India have been unfolding and the chorus of support and opposition keeps mounting, political events have taken an unprecedented turn. For a time it seemed that the BJP was not only the main catalyst but was also fated to be the 'innovator' to whose innovations the rest of the parties would have to respond. The BJP became the main opposition party in Parliament and came to power in Rajasthan, Uttar Pradesh, Madhya Pradesh and Himachal Pradesh. The impact of this has been felt in new alignments with the Left in co-operation with breakaway parties of alienated Congressmen like the National Front parties including the Janata Dal, veering towards the secularism of the Congress while warning it to oppose communalism and corruption in politics. This emerging coalition, where power fluctuates according to the electoral strength of each group is opposed by the BJP whose grassroots support, particularly among Hindi-speaking Hindus, is galvanized by symbolic events like the Ayodhya episode, through which the BJP is popularly perceived as the protector of India's Hindu heritage and future in a world that fears the spread of fundamentalism. Indeed, widespread migration into the Eastern region of India from Bangladesh evokes fears of a movement for *Lebensraum*. The noted BJP ideologue, Govindacharya, told this author: 'In India, and for that matter elsewhere, the debate on secularism will remain inconclusive unless one discusses threadbare the approach of Islam to non-believers ...'. At the same time he adds, 'since the quintessence of Indian nationalism is *Hindutva* — and the first requisite of *Hindutva* is respect for all modes of worship — and Hindu Rashtra is not synonymous with Hindu State, freedom of mode of worship will be duly protected and promoted'.

Later events, however, exposed the heterogeneity of the ideological, caste and class composition of the mass bases whom the BJP wished to mould into a homogeneous 'Hindu nation'. The issue of reservation of jobs for the so-called backward castes came to the fore in North Indian politics. In Madhya Pradesh and Himachal Pradesh the Congress swept back to power, and in Uttar Pradesh, a coalition between the Bahujan Samaj Party and the Samajwadi Janata Party defeated both the BJP and Congress decisively in the polls.

Some secularists and scholars on the left believe that with these challenges at the class and caste levels, the ideological basis of the 'secular' claims for Hinduism made by the BJP and those in quest of a united Indian nation remains infirm. Amiya Bagchi stresses: 'both the Mitakshara and Dayabhaga Schools of Hindu Law are rooted specifically in the Hindu view of the world and in that respect are much more analogous to Muslim Personal Law than most Hindus realize; all of this should be reformed to align them with a concept of civil law which has no truck with religion. Such a civil law can be joined with company and tax law so as to make for better tax governance and better civil contracts'. Internationalists like Amartya Sen, proclaim the need to protect and maintain India's pluralism, and humanists like V.S. Naipaul speak of 'the state of the country: destitute, trampled upon, crushed. You then had the Bengal Renaissance, the thinkers of the nineteenth century. But all this came from the top. What is happening now is different. The movement now is from below.' He adds: '... we all live in a universal civilization, some more than others. We have individual particularities. But we are all inhabited by a universal civilization. It is very hard to go back.'

Although economic reforms have been undertaken and religion has been brought to the forefront of Indian politics, debates about what has been done and what the Indian people might want to accomplish are as fierce as ever. These debates have generated a rethinking of the basis of Indian nationalism in the period before

Independence and after. Some of the authors who are represented in this volume have already contributed fruitfully to the debate.[1]

While liberalization has proceeded at a fast pace according to most observers in India, the political success of these measures has not silenced voices which stress the continued relevance of state intervention, public enterprises, and economic nationalism for the fulfilment of the objectives by these reforms.[2] On the other hand, there are strong supporters of economic reforms who believe that they have not proceeded as fast as they would have liked.[3]

CONCLUDING REMARKS

In the essays collected in this volume, the authors reflect and advance their views of the transition process and their special visions of the India they would like to live in and leave behind for future generations. The editor has made no effort to either iron out the diversities or to impose a uniform format on the area or style of discourse. All of the essays are concerned with the present. But theoretically the present is always an evanescent, ever-moving point of infinitely small dimensions between the receding past and a looming future. The memories of the past and visions of the future are almost inevitably imprecated in the construction of the present. How those memories are retrieved, how the recent past is grieved over or celebrated, and how the future is captured in the mind's vision provides its special flavour to committed political discourse. The editor can only present the discourses as faithfully as he can, and should not try to order their life process if they are to resonate in the mind of the reader.

1 P.R. Brass: *The Politics of India since Independence* (Cambridge University Press 1990); P. Chatterjee: *Nation and its Fragments* (OUP 1994)

2 T.J. Byers (ed): *The State and Development Planning in India.*

3 Ashok Desai: *No Economic Affair* (Wiley-Eastern, New Delhi 1994)

India in Transition: An Overview

PRADIP SARBADHIKARI

In the half-century since Independence, India's policy-makers, party leaders and intelligentsia have offered a variety of alternative approaches for her national development. These approaches have focused on four major themes: the State, the economy, social change, and India's place in the global system. While there are a bewildering number of approaches to these four central themes, a typology may be suggested based on ideology and modes of thought. These can be divided not only into conservative and radical policies as they pertain to say, economic and *social* policies, but also tradition and modernity as the his introduction, these alternative and contending perspectives on the four central themes have been enuncy apply to the role and philosophical basis of the State.

In the articles that follow, these alternative and contending perspectives on the four central themes have been enunciated by eminent scholars, party leaders and cabinet ministers. The objective is to understand the diverse and changing concepts of the Indian State: the opportunities and pitfalls of economic restructuring, of values in promoting social reforms, of ensuring security in a new world order, as well as to evaluate the success and failure of policies. My main purpose in these pages is to provide an overview of some salient features in India's transition in order to provoke and stimulate discussion rather than provide information. As a general methodological observation it would seem to me that scholarship on the

Indian State, economy and society has lent itself to an overriding, though not exclusive, economic interpretation. This is largely understood in the context of India's economic and social history rooted as it is in her colonial and pre-colonial past. However, such an explanation cannot deal with all the modern complexities and remains incomplete; hence, I would extend this to agree with Joan Robinson that there are 'below it layers of geography, biology and psychology and in layers above it the investigation of social and political relationships and the history of culture, law and religion'.[1]

DIMENSIONS OF CHANGE

By 1947 India could lay claim to belonging to both the traditional and modern worlds. In the half-century or so since then, the impulse to modernize has been countered by the will to retain traditional values, institutions and cultures. These contradictory forces have centred around political movements and policy changes to promote political freedom, economic growth and social progress. The wide array of choices amongst the processes to be followed to achieve these objectives has given rise to differences both amongst Indian and foreign scholars. These debates in turn have been modified by the impact of the policies on the lives of different sectors amongst the people, resulting in a continuous shift in policy. While these shifts have occurred without major conceptual and structural changes resulting in a socio-economic transformation, the current national political and economic changes and reshaping of the international structure have created a new watershed in her history. India's policy-makers have started boldly responding to these changes setting in motion new debates, some of the leading protagonists of which are represented in this book.

Earlier, the debate on political freedom and constitutional change had centred on the merits of a Westminster-style Parliamentary democracy with its Presidential modification, the strength and weakness of one-party people's democracies, partyless communitarian politics and more recently a call for a National Government made up of all parties.

It is fair to say that at the formal level, the option of a people's

1 Joan Robinson, *Freedom and Necessity*, 1970

democracy was never seen as a viable alternative. Class, ethical objections, comparative experiences (e.g. China's Cultural Revolution and Cambodia's de-urbanization), institutional (e.g. the police and military apparatus) and other structural considerations, preempted any formalization of informal access which advocates of people's democracies may have obtained in the rural hinterland through liberation struggles. Likewise, the partyless communitarian politics with its characteristic non violence and back-to-the-village impetus was never regarded as a legitimate and rational alternative, despite its heritage in Mahatma Gandhi's philosophy and Jayaprakash Narayan's active espousal over a lifetime. The adoption of Panchayat Raj and village democracy would be under the aegis of a Parliamentary system and not its substitute.

The models of rural revolutionaries adhering either to Marx or to Gandhi had been excluded for their utopian and anarchist ideas whether sought to be communicated through violence or non-violence. Consequently, the political alternatives have been circumscribed by agreement on the form of a parliamentary democratic system which, despite its shortcomings, has been legitimized by world public opinion.

In contrast to the rural-centred alternatives, the Westminster Parliamentary system was the most familiar with the nationalist and post-nationalist leadership and easily filled the vacuum left behind by a receding British Empire. The constitution, with intermittent amendments, would be modified to fit in with objectives set out by the leadership even and/or precisely because it would strengthen the power of the State even at the cost of human rights. Admittedly, the necessity of the State to protect its citizens against violent acts of terror does not need stressing yet there is little doubt that economic and social compulsions would be used as an excuse for 'collective rights'. Individual civil rights though constitutionally preeminent and vigorously safeguarded would, barring exceptional cases, become a dead letter in the face of the Indian State's transgressions of human rights and the political and social elite's blindness, apathy or silence. An archaic judicial system, in which individual access is more a fiction than reality, coupled with an insensitive and parasitic bureaucracy would aid and abet these transgressions.

More recently another alternative model, also within the Westminster Parliamentary tradition, is being offered albeit on the premise of an ethical and just collective leadership. Such an inter-party National Government is however seen by its critics as little more than a formidable bargaining chip that does little to assuage regional and sectional needs.

India's parliamentary democracy has remained a formal institutional framework which is unique given the history of parliamentary democracies. The Puritan Revolution, the French Revolution and the American Civil War gave Britain, France and the USA bourgeois revolutions respectively on which they could build a system of liberal democracies with *de jure* individual rights enshrined in their constitutions. The Soviet and Chinese revolutions with strong peasant and peasant-workers' revolutions also created the conditions on which civic freedoms could have been built; it is the failure of these expectations to legitimate consent that has led to radical restructuring in the former but continued coercion in the latter where civil freedoms have been delayed. A third set of states like Germany and Japan and, to a lesser degree, Spain and Italy, followed a capitalist route and came to fascism and its aftermath, despite which industrial growth and economic advancement occurred, and created conditions for the present liberal democratic regimes. India, in contrast, has had no revolutionary history, neither bourgeois nor communist, despite the sporadic radical uprisings and social and political movements during the colonial and post-colonial periods; nor has it through the totalitarianism of colonialism or authoritarianism of the post-colonial period been able to develop a flourishing industrial surplus from which she could draw the life blood of individual rights and freedoms although these are ritualistically mentioned like mantras in the Constitution and Court systems. India's situation is unique and there are serious contradictions — the constitutional framework for parliamentary democracy and economic stagnation if not the 'development of underdevelopment'. If parliamentary democracy is the path that national parties of the Left and Right see as the political model for twenty-first century India, its economic complement through rapid industrialization, land reform and other variables of economic change is urgently needed.

The consensus about constitutional government and a parliamentary system does not however minimize political conflicts stemming from the politics of recognition which is intertwined with the politics of humiliation and despair. Economic, religious, ethnic, linguistic, caste and now the faint voice of women's rights are being articulated in many forms: from regionalism, to secession, to the modernization of individualism, requiring new federated, confederal or socio-economic structures, institutions and most of all an attitudinal change that grows out of reinterpretation of culture and tradition.

Complementing demands for these new political relationships are divisive economic goals and processes. The grand debate between greater encouragement to the private sector and of State control has been waged in the foreground of a larger issue of economic self-reliance versus liberalized policies in a global trading regime as the Brandt Commission Report and its critique had foreshadowed two decades ago.

Liberalization has had a close theoretical and policy linkage with modernization. Modernization as a social theory has had a rich intellectual tradition in the post-war years. In the fifties and early sixties theories of modernization stimulated much debate on the strategies of development. With the Vietnam War and its linkages with theories of modernization, the practice of U.S. military intervention cast its shadow on modernization. The field had meanwhile been appropriated by other theories, notably the dependency theories initiated by Latin American scholars and adopted by other scholars of the Third World. It took ten more years, during which the limitations of dependency and its variants especially in formulating policy became more apparent, for modernization to receive renewed scholarly attention. China's return to the comity of nations was as pertinent a factor as Cambodia's ghastly failure with isolationism, as well as the contrasting success of other S E Asian states which had successfully opened their economies to investment and trade under the hegemony of the State.

The recent repudiation of 'self-reliance' through exclusive State planning, exemplified in the collapse of socialist governments, has given new grist to the mill. The irony however is that the crisis of socialism cannot blind the dispassionate scholar to the deepening crisis of capitalism reflected in sustained recession, unemployment, and budgetary deficits and accompanying social changes. It is in the context of the balance of these forces that the situation in India must be viewed. Indeed, the lure of the market has fired the imagination of political leaders across the world, leading to a scramble for foreign investment and technology which has strengthened the hand of multinationals, banks and monetary agencies immeasurably. Such enthusiasm among those who have gone to the market could have been a refreshing change from the dogmatism and stagnation of the past but for the necessity to moderate economic changes so that the economic deluge from perestroika is not reflected in the rest of the world. It is this debate inaugurated in India by drastic economic reforms of the new Government that has renewed the promise of economic regeneration and the nightmare of loss of Swaraj. In any event, given the high velocity of forces of changes across the world, India cannot stand still, and however it turns out in the future this is indeed another turning point in its history.

THE INDIAN STATE AT THE TIME OF INDEPENDENCE AND AFTER

A discussion of change in contemporary India has to take into account the inclination and the will to reform which have provided the wellsprings of contemporary Indian history. At times this will has remained dormant, or not manifested itself in actual policies and movements, due to external and internal constraints. On other occasions the will to reform has led to fundamental shifts in India's historical process. The nationalist awakenings in the late eighteenth and nineteenth centuries and the social and political movements of the first half of the twentieth century followed from these zealous initiatives for change.

In the aftermath of Independence, the change from a colony into a sovereign State with new equations of political and economic power, placed strains on the Indian polity which were not always perceptible in the euphoria of freedom. The emphasis of its new leaders was on a powerful centralized State with a bureaucracy that had been loyal to the British Raj but was inducted because of its experience and administrative competence. This new State was dressed in the formal regalia of a federal State whose writ ran across divergent castes, linguistic, religious, ethnic and regional communities. The new authority was sought to be tempered by the consent of the population through universal suffrage but regulated by a *de facto* one-party system. Over the decades since independence this uniparty mould would begin to loosen up and it became increasingly evident that the party system at the provincial or sub-national level would interact with the social, cultural and economic identities and disparities, and lead to the introduction of a multiparty system that would challenge the informal premises on which the State was created in 1947.

Such political changes have proved to be superficial, even cosmetic. The acquisition of political sovereignty after a long and complex nationalist movement, often heightened by a bitter and heroic patriotic struggle, did not unfortunately inaugurate a socio-economic transformation. Despite sectoral improvements reflected in statistical changes and the strengthening of an elite drawn from the educated or political party leaders and patrons, the vast majority of the population lay mired in colossal poverty and backwardness.

When Britain left India, the new nationalist State was created under conditions of formal political freedom. Since there was not one homogenous India, the interests of all Indians were not felt equally. There was not only a great divide between religion (which had inexorably or unwittingly led to Partition), but also rivalries of caste, ethnolinguistic groups, and regions which transcended class divisions. The new political edifice that emerged provided only limited access for individual upward mobility through severely constrained bureaucratic and economic channels. Yet the contradiction, between a new Indian State proclaiming democracy and social justice, and a coercive State apparatus in which an administrative and economic elite often drawn from landowning classes that profited from the resources of the State, became sharper. This contradiction filtered into the social schisms leading to inter-caste, inter-religious or communal, linguistic and regional conflicts often, though not always, identified with economic and class differences. The policy of linguistic States, like other policies in response to crises, was politically motivated and this may indeed have exacerbated low-level tensions. The conflicts that resulted have so far been suppressed by the might of the State or mediated by political, economic and social processes and confined to regions. Whether the pressures that have been building up within the polity, often articulated by individual and collective violence, will lead to a national explosion or regional breakdowns as we have recently witnessed, should become evident as India enters the new century.

STATE POWER: A COMPARATIVE PERSPECTIVE

Has the State become too powerful? This has been a fundamental question first raised in the West since the original euphoria of victory evaporated in the post-war years and the new Europe adjusted itself from its post-Empire to European Community status. It was the growing debate about the reaches and limits of the welfare State and its weakness in responding to new economic and social pressures, often induced by technological innovations, that occasioned the decline of social democratic regimes and the parties that had fuelled them — notably the Labour Party in Britain, the Social Democratic Party in Germany and the Left in Italy, France, and Spain which had cloaked itself in the ambiguity of Eurocommunism and a mixed economy socialism.

Even in Scandinavia, where the welfare State had become sacred, there is growing disillusionment manifested in the challenge to Social Democratic Parties in Sweden, Norway and Finland. Across the Atlantic, the Reaganite programme was a continuation of the Republican Party's onslaught which had been aborted at the time of Barry Goldwater's Presidential challenge and quietly began with the Nixon Presidency; Jimmy Carter despite his close relations with the Trilateral Commission, provided a short Democratic interregnum and a last echo to Lyndon Johnson's Great Society. In the eighties it swept away the social gains made since the public acceptance of Roosevelt's New Deal philosophy. With the Clinton Presidency there has been an initial attempt to redress the balance in favour of social issues, particularly health insurance, which the Presidency was forced to abandon. Critics espousing the liberalism of the Democratic Party say that the appeasement of the right through abandonment of emphasis on social and economic policies and posturing, as in foreign policy escapades into Somalia and Haiti, turned out to be fruitless. The control of both Houses of Congress by Republicans led by a fundamentalist Right in midterm elections at the end of 1994 has underlined the American public's disenchantment with the tradition of Roosevelt and Kennedy-Johnson's Great Society programmes. These, they argue, have been overtaken by massive budgetary deficits, industrial unemployment, a polarization of incomes, breakdown of civil society and the family. This formal shift to the Right will be reflected in America's policies in the world arena. But its impact has to be limited since the new world order is already different from the one expounded by President Bush; as scholars begin to predict: the most spectacular component of the current revolution is 'the shift in the world economy's centre of gravity to Asia. ...... Indeed, if present trends continue, East Asia should be poised to claim pre-eminence over its two counterparts (Western Europe and North America) before the turn of the century... ...'.[2]

Supply-side economics even made inroads into Ottawa and Mexico, opening them up to the North American Free Trade Agreement. The cumulative effect of these political changes which originated from growing concern with the power of the State, and concomitant decline of individual initiative and enterprise loudly proclaimed by Thatcherism, was to destroy the citadels of the State system, first in Eastern Europe and then in its ideological bastion in the Soviet Union.

While the historical experience, idioms of change, and national development have been significantly different in the Third World, the rumblings of change in the West have had their impact on the erstwhile Third World leading first to whispers and then increasingly to strident voices for change in the *status quo* in the Third World. These challenges have been manifested in the decline of military regimes in Latin America and the defensiveness of authoritarian regimes in Africa. In Asia the pattern has been more ambiguous although the events in India are characteristic of a steady erosion of Central-State authority through demands for fiscal and economic independence in a new federalism, and linguistic, ethnocultural self-determination.

This trend towards decentralization and political devolution has usually been understood in the context of Western politics, whether in Scotland and Wales in the UK, the Flemish-speaking in Belgium, Brittany in France, or Quebec in Canada. In post-communist States of Eastern Europe and the Soviet Union, the forces of self-determination have taken the ethnolinguistic map back to the years before the Versailles Treaty and the Russian Revolution. At the political level it reflects the global transition from centralized State power to

2 Klaus Schwab and Claude Smajda: 'Power and Policy' in *Harvard Business Review*, November-December 1994, pp. 40-50.

regions, provinces and local communities, though it is impervious to the corporatist multinationals: such is the need for capital and technology for investment and re-employment. But centralization, and devolution and self-determination, do not explain the trends in religious revivalism. Social psychological explanations complementing those of the history of culture and religion have been offered to account for religious revivalism most prominently in the Islamic States and now for a reassertion of Hinduism and militancy of Sikhism. The West too, has not been unaffected by the rising theocracy, as the New Right and its Evangelical collaborators on television seek to destroy the long tradition of separating Church from State.

LEADERSHIP AND INSTITUTIONS

Charismatic leadership had galvanized nationalist movements in the colonial States. It first featured prominently in late nineteenth-century and twentieth-century India and played a pivotal role leading to independence. It also inspired other colonial States in Asia and Africa. While Mahatma Gandhi, Netaji Subhas Chandra Bose and Jawaharlal Nehru formed a pre-eminent Trinity (TRIMURTI) in India's nationalist lore, their roles were matched in Asia and Africa by Aung San, Sukarno, Nkruma, Sekou Toure, Kenyatta, Ho Chi Minh, Ben Bella, Mao Zedong, Nasser, Bourguiba, Mandela and other leaders in their respective nationalist movements.

The transition of the nationalist parties to ruling parties of new sovereign States while legitimizing the charismatic leaders as heads of government, underlined the importance of institutions in the new political process. Some leaders in their wisdom adjusted to the new structure; others who ignored the autonomy of institutions did so at their peril. The contrast between Jawaharlal Nehru and Indira Gandhi or between Nehru and Nkruma exemplifies this choice. In India, the Nehruvian phase stretching to the early/mid-sixties saw a conscious effort by the leadership to utilize the party structure to its own advantage while explicitly or implicitly recognizing the party's particularistic regional autonomy. The Congress Party despite occasional exceptions continued to remain the dominant party both at the Centre and in the States.

This carefully worked out doctrine of institutional government, albeit with a charismatic leader as the supreme arbiter, continued beyond Nehru's life. Lal Bahadur Shastri played a pointedly low-key political though effective administrative role. The succession of Indira Gandhi and inauguration of what has since been called a dynastic system would obviate the necessity of institutional autonomy. Hence the Congress Party began to lose its autonomy in the States while power was centralized in the Prime Minister and his friends. The imposition of the Emergency on personal grounds identified the State with the leadership and epitomized this breakdown in the democratic system. Non-Congress Governments headed by Morarji Desai and Charan Singh provided brief interregnums marked by personal ambitions, political cabals and disarray. Rajiv Gandhi, says James Manor, 'achieved a kind of transcendence' during his Prime Ministership. It has been hotly debated whether this can be construed as moral leadership, political vision, or dynastic and partly legitimacy. Whatever the interpretation, such 'transcendence' evaded both V.P. Singh and certainly Chandrasekar whose first priorities were to keep their minority governments in power. Hence it is to the credit of Narasimha Rao that while seeking an accommodation with the opposition he has been able to lead the Congress at the Centre despite its weakness in several States. This success, however, will have to be matched by strong organizational support from factions within the Congress including successors of the Nehru-Gandhi political heritage symbolized by Sonia Gandhi if it is not to wither away in the next decade. Such creative politics must inevitably come to terms with the principles of parliamentary government and build a civil society on a national scale if India is to survive as a constitutional multiparty system.[3]

Notwithstanding the shocks to India's democratic process, it must be pointed out, that the unprecedented increase in the power of the Head of Government was not confined to Mrs Gandhi and to the Nehru family. In other Westminister model States the same process has been increasingly apparent. In Britain and Canada there has been an increase in the power of the Prime Minister's Office (PMO) in inverse ratio to the decline in the power of the cabinet and responsibility to the Party. The old dictum *primus inter pares* (first among equals) has been replaced by *'L'Etat c'est moi'* (I am

3 James Manor (ed.) *Introduction to the Nineties*, UBC Press, Vancouver, 1994.

the State). The centralization is more acute in non-parliamentary systems.

Several explanations have been offered for this new trend and erosion of institutional autonomy in the political process; they range from politics in the technotronic age to national economic dependency in a global trade and investment regime, to political alienation based on language, ethnicity and race. These reasons however convincing do not justify, especially when the democratic system is invoked, the steady erosion of responsible government at whose core is the structure of autonomous institutions including a representative party system. If such strength is vitiated by interest group-politics that is symbolic and ritualistic — seen as vehicles for letting off steam — such processes need to be corrected rather than abandoned. In India, political pluralism also needs to be rescued from bureaucratic control. It is a moot point whether the threat to Indian democracy is greater from an unresponsive, deadening bureaucracy or from ambitious, even predatory, political leaders and their cohorts.

According to Pranab Bardhan, the heterogeneity and plurality of the proprietary classes has contributed to the preservation of the democratic processes which constitute a permanent feature of the Indian polity. Nevertheless the trend towards a personalized political structure is evident in the current phase of Indian politics and has given rise to speculation about the setting up of a Presidential system of government.

The debate does not have to be confined to dynasty-control versus democratization as some of the changes made by Rajiv Gandhi before his tragic end would suggest to future historians. Yet, to a charismatic dynastic leader the process of democratization, however enlightened as an objective, is an immensely formidable process. Neither the political parties nor functions within them are endowed with moral virtue on this question. The onus lies on the leadership and middle-level party workers to genuinely reflect the will of their members in order to play an effective intermediary role between the State on the one hand and a highly concentrated economic power and society on the other.

ECONOMIC GROWTH AND RESPONSIBLE GOVERNMENT

The case for a parliamentary system based on consent has other detractors. Some scholars, notably developmentalists, have argued that desperate needs of economic growth in underdeveloped States have such priority that representative governments and autonomous party systems could become not only an unnecessary luxury but an obstacle to growth. Such a view, ironically shared both by advocates of stabilization and structural adjustment as well as some varieties of dependency theories, has its logic, particularly in the short run, when radical populism can be mobilized for individuals to make sacrifices for promoting the social good. But an alternative argument backed by empirical evidence makes a strong case for sustained long-term economic growth, whether in Africa, Latin America, Asia or for that matter in Eastern Europe. Recent changes have indeed highlighted the importance of individual consent articulated through a representative party system as the *sine qua non* of individual initiative and enterprise as an essential source of national economic growth. The jury on these two differing political conditions for economic development are still out. It needs to be added however that there are contradictions indicating consent as only one of several variables for economic development. Theories of modernization based on cloning Western market systems, and social structures notwithstanding, one cannot help recognizing conflicting results from statist regimes of Africa and Latin America on the one hand and corporatist States like Japan, South Korea, Singapore and the other newly industrializing countries on the other. However, India's quest to devise effective political conditions for successful developmental strategies based on a mixed planned economy, and parliamentary rituals with charismatic and dynastic leaders in *de facto* one-party systems whose regional autonomy has been constricted, has not brought much relief.

Dr Pranab Bardhan[4] has also raised the question why, in spite of having enormous resources and an impressive administrative and industrial infrastructure, India has not developed into an economic giant. What is astonishing is that over 30 per cent of the people still live below the poverty line. One reason is the cut in social welfare

4 Bardhan: *The Political Economy of Development in India* (OUP, 1984)

programmes for the poor. Since the poor are not an organized social group leakages of goods and benefits occur, as a result of which these do not reach those targeted. The high growth rate in some areas is diluted by a low growth rate in other areas. The State has played a major role in development in the last few decades but India lags far behind some East-Asian countries, South Korea for instance. Dr Bardhan has given the reason in one sentence: 'In spite of the impressive record of resource mobilization, largely through indirect taxation and transfer of savings from the household sector to nationalized financial institutions, the bulk of the resources are frittered away in current expenditure, with the result that not enough surplus is left over to finance massive public investment programmes in such areas as coal, transport, power and irrigation.' Furthermore, because of the plurality of proprietary classes — the industrialist capitalist class, rich farmers and professionals in the public sector — pulling in different directions, there has been a proliferation of grants and subsidies which has slowed down development.

It is in this backdrop that efforts to restructure the economy, reform the political structure, and reconstitute Indian society have to be viewed. Such a view must, as a matter not only of prudence but of necessity as well, question the motives and wisdom of reformers and their reforms given half a century of hopes, frustrations and despair. But its critique must not be motivated by the protection of vested interests, whether they are economic, political or administrative.

THE INTERNATIONAL LEFT AND THE INDIAN STATE

The nineties are too early and not the most propitious years for analyzing the success and failure either of Marxist theory or of Communist Governments' practices dispassionately, logically or conclusively. The historic collapse of Communist Governments in Eastern Europe followed by that of the Soviet Communist Party and disintegration of the Soviet Union under Gorbachev's perestroika suggests a chillingly incongruous domino effect — a ghastly mocking echo from the Vietnam of General Giap and Ho Chi Minh. The economic rubble left by political changes may contain the seeds of individual enterprise which the statism of earlier decades had buried; it may also generate new political forces which accord due reverence to the humanism which inspired Marx against the tyranny of states and societies in nineteenth-century Europe. Despite these ironies of history it is very pertinent to examine the status of the Indian Left as it becomes disconnected, even alienated, from the social democratic Left and then from international communism, notwithstanding its new fraternity with a post-Tiananmen Square Chinese regime. While communism as a political force has as long a history in India as in China, and Indian Marxists like M. N. Roy were the fountainhead of the international Left's anti-colonial revolution, Indian communism has not achieved the euphoric success which Lenin's revolution gave to the erstwhile Soviet Union, Mao Zedong's Long March gave to China, or Ho Chi Minh and 'Che' Guevara's leadership gave to the Vietnamese and Cuban revolutions. Its success has been limited to party victories won through the ballot box in West Bengal and Kerala, inviting charges of deviation towards social democracy. Perhaps a more unequivocal measure of success in the context of conventional Marxism is in the Kisan Movement, in the trade union movement, and in those villages which had come under the hegemony of radical rural regimes in Naxalbari, Andhra Pradesh and Northern Bihar.

The reasons for this rather odd historical record have been sought in a variety of explanations that range from the incompatibility of an atheistic, rather than simply secular movement in a civilization that is both spiritual and fatalistic, to political strategies that ignored popular sentiments either by supporting Britain's war efforts or condoning the Emergency. It is only lately that with the CPI's identity crisis and the CPM's new-found strength in a multiparty national political system, buoyed by skilful leadership and dedicated workers, that communism in India has won, however fleetingly, a unique status unhindered by the disasters of the international Left. But in spite of the nationalization of communism in India the continued obeisance to the high priests of Communist orthodoxy, especially to Stalin, casts a shadow on the humanistic aspirations of the movement. It has alienated many with a socialist consciousness. Likewise, the reiteration of economic dogma based on the infallibility of State enterprise, even in a globalized economy characterized by technological revolutions and capital movements, needs courageous rethinking if the Left in India is to address the problems that have dehumanized the masses. Despite these reservations, and the urgent need of the party to convert organizational skills into efficiency and honesty in administration, and prevent the nightmare that urbanized sprawls have become, the role of the Left in ensuring communal harmony, as in Bengal, has given it a historic national status. Informal coalition politics has set the stage for creative democratization.

TERRITORIAL NATION STATE OR CULTURAL NATIONALITY

Alongside the debate about federalism and further decentralization is a new debate about the identity of the Indian State. This debate has grown louder and shriller as the new national political parties like the Bharatiya Janata Party (BJP) have assumed greater power not only at the national level through increasing electoral strength in the two Houses of Parliament but also in the State Assemblies. While not denigrating Nehru personally, the new debate seeks greater scrutiny of the Nehruvian model — secularist, social democratic, mixed economy, non-aligned State — and found it to be dated at best or at least severely wanting. Nehru's critics have found his social policies to have been gravely flawed with abysmal conditions in primary education, health and nutrition, language policy, family planning, and in a culture of intolerance: his secularism they have alleged had 'degenerated into equal respect for everybody's communalism' and a foreign policy 'unable to comprehend, let alone deliver, what India needs'. The critique of Nehru's policies and his legacy has led to sharply differentiated conceptions of India's Statehood. While the past leaders, notably Nehru, articulated a Western conception of the Nation State to provide territorial sovereignty, security, and national development, the alternative models develop an extraterritorial, cultural, even a civilizational view and thereby an inherently theocratic concept of the Indian state. Here the objective is to revive tradition and find the future in the past or fuse tradition with modernity. Such a conception does not necessarily rectify the Nehruvian models' weakness in achieving a balance between a State borrowed from Westminster and society that is predominantly religious,ritualistic and remains pre-industrial. Whether this cleavage can be removed by a civilizational concept of the Indian State or would remain shrouded by it, while exacerbating the economic and social problems that modern States in an integrated global world have of necessity to contend with, is a direction the debate could usefully take. In any event, such a debate itself is intellectually enriching as it dusts the mantras of non-alignment, secularism and socialism and seeks to deal with the constitutional as well as conceptual basis of the Indian State.

Francine R. Frankel has made a useful analysis of how caste, class, ethnicity and dominance are treated as structures and processes interacting with each other and with increasingly powerful State institutions; how state policies undermine the religious legitimacy of the hierarchical social order and facilitate manipulation of linguistic, communal, caste and ethnic loyalties; the results of a diffusion of class polarizations: the battle for ascendancy between city and countryside, industry and agriculture, bureaucrats and politicians, urban middle class and peasantry, landowners and agricultural labourers, industrialists and workers, forward castes and backward castes, and even between advocates of a Hindu State and those of a secular polity. These she argues are aspects of a social turmoil accompanying the collapse of a Brahmanical system as the source of legitimacy for the unequal social order.

ECONOMIC PARAMETERS: BEYOND STATE HEGEMONY

Political freedom in 1947 was have been complemented by economic freedom, not only measured in national terms of self-reliance and growth but also in terms of individual economic security. The primary vehicles for this were the succession of Five Year Plans which have sought national development under various levels of State hegemony: primarily public sector enterprises.

The Left critique was that despite the moral origins and noble goals, economic planning was directed to the promotion of dependent capitalism in India; public sector enterprises were in fact the preserve of bureaucratic elites and party functionaries and essentially vehicles for siphoning off peoples's resources for private and/or corporate gain. The joint sector enterprise between State and private capital was no better and often abused as a mechanism to legitimate the flow of State funds into corporate or bur ucratic hands rather than encourage and strengthen small- and medium-scale investors.

The critics from the Right chided the policy-makers for

promoting policies that strengthened the inefficiency and corruption of bureaucrats and political leaders under the mask of self-reliance. Most of all, the ideological idiom-permeating criticism of the Government's policies from the early Swatantra Party and Jan Sangh to the present Bharatiya Janata Party (BJP) was that State control discourages individual enterprise which is crucial for economic efficiency, India's productivity, growth, and competitiveness in a global economy. This ideological view has at times been obscured by the division within the Right between populist measures embracing nationalist postures and capitalism, both national and multinational.

Meanwhile, throughout the eighties, the international economic environment was becoming charged with supply-side economics under the inspiration of Reagan and Thatcher. The world was also overawed by the spectacular success of Japan and the newly industrializing States, whose achievements contrasted remarkably with those of Latin American and African States, and more recently the former socialist States of Eastern Europe, pursuing policies of self-reliance.

These international economic changes were complemented by the weakening of India's *de facto* one-party politics leading to new alignments and the contours of coalition politics. The US war in Iraq created a further foreign exchange crisis exacerbated by higher oil prices and sharply reduced remittances from non-resident Indians which brought matters to a head occasioning or providing a rationale for the new economic reforms.

One of India's recent critics touching on its present economic policies is Moody's which has acted as a barometer in international finance. It argued that India's aid bill had increased partly due to the increase in debts leading to higher debt-servicing costs since commercial borrowing had also increased. The inflationary pressures and excess demand coupled with increasing external and fiscal imbalances all contributed to a new economic crisis since 1987. When the 1990-1 budget was presented, Moody's declared that 'it believes that the Government does not have the capacity to achieve a rapid improvement in the Government's budget deficit: it has little discretion over a large chunk of the budget because of the nature of the expenditure and India's fractious domestic political conditions make it difficult to cut spending that sustains the Governments' political support'. Like Moody's, other international institutions' perceptions of India have been negative. The OECD in a report on Financing and External Debt of Developing Countries concluded that India's debt had more than doubled between 1982-5 and 1986-9.

It was now a matter of time before the Finance Minister would introduce changes in industrial and trade policies; further loans would lead to spiralling of her debts, and stabilization and structural adjustment policies would be urged and imposed under the aegis of the World Bank and the International Monetary Fund. As in other countries which have experienced these policies, notably in Africa and Latin America, the new reforms foreshadow continued fiscal policy-changes and changes in exchange rates; an expanding private sector operating in a competitive global marketplace; smaller size of governments necessitating reduction of payrolls; foreign investment, including those in previously nationalized sectors, inevitably by multinational corporations; spending cuts; elimination of price controls; and tariff reductions (John Williamson, 1990). The impact of these became evident as the Government began to implement these policies when there began a decisive shift from State power to market policy.

LIBERALIZATION AS A SELL-OUT

The new policies led to a storm of protest. Despite the attraction to overseas investors, conditional dividend repatriation based on export performance is likely, it was argued, to limit overseas investment. And although multinationals are sought for using India as a base for exporting to the rest of the world, their interest might in fact be in India's huge domestic market rather than promoting India's exports as the new industrial policy beckons. Besides, multinationals have several alternative locations to choose from because of competition to attract scarce overseas capital. Such locations competitive to India are in Latin America, selected African States, and more prominently in Eastern Europe where the disintegration of the State machinery has created an economic vacuum similar to that in Western Europe on the eve of the Marshall Plan.

Nevertherless, the World Bank's loans are a serious threat to India's economic independence and national development. Cheryl Payer prophesied that 'the Bank's loans to India in 1981 were on easy terms as India boasted, but the screws are gradually tightening ... the Bank is demanding devaluation and much more borrowing from commercial banks'. Ironically Dr Manmohan Singh, now Finance Minister, as Secretary of the South Commission in 1988 expressed 'grave doubts concerning the theoretical validity of some of the key prescriptions now involved in conditionality. Financing liberalization in conditions of inflation has led to aggravation of inflation. Insistence on import liberalization in periods of pressure has led to aggravation of the balance of payments deficits and frequently to devaluation to a greater degree than would be needed.'

Critics of devaluation say it will not lead to a narrowing of the trade gap because apart from the dismal experience of Africa where case studies reveal a sharp increase in the unofficial value of the dollar, much of India's export in primary and low processed goods are usually fixed in dollars; other exports like garments are regulated by volume, and price is not a limiting factor. Devaluation is seen by these critics more as a dogma rather than an effective economic measure. It does not, they argue, address the problems plaguing India's public sector industries, such as political interference in management, administered prices, bad technological choices and inefficiency.

The debate on conditionality has been overtaken by the debate about the GATT provisions, and vociferous opponents on the Uruguay Rounds final treaty have been matched by free marketers who received a boost from the creation of the World Trade Organization in 1995 which would embody and implement the GATT treaty. Meanwhile, India has been receiving unusual attention from the Western media and multinationals which has led to an influx of international capital, making opponents so much the more wary of changes in domestic legislation. Whatever the outcome, these economic events signal the beginnings of a new era in post-independent India's economic history.

FOREIGN POLICY

Non-alignment and National Interest: from Ritual to Interests

Two issues about the effectiveness of India's foreign policy deserve special comment. First, the continued 'ritual' of non-alignment and second, the invocation of an ideological foreign policy rather than one explicitly based on capability, geopolitics and national interests.

The foreign policy of non-alignment reflected the ideology of Krishna Menon who conceived of it and was its most articulate spokesman especially at the United Nations: Jawaharlal Nehru, by promoting the policy, enabled it to become India's most notable contribution to the post-war era. Afro-Asian leaders who had earlier been inspired by India's nationalist movement, which opened their door to decolonization, quickly adopted India's non-aligned policy for their newly sovereign States; even Latin America and independent-minded European leaders like Tito, struggling to free themselves of hegemonic shadows, extended non-alignment into an international doctrine.

Despite this success, what requires serious re-examination at this watershed of global change is whether the policy apart from raising India's diplomatic and international 'status' contributed towards her national interests. The record on this is, of course, ambiguous: it certainly embarrassed India during its conflict with China when non-aligned States remained aloof in the conflicts between the Great Powers of Asia, and India felt 'betrayed' over Tibet and borders with People's China after having strenuously championed the latter's admission into the UN for over a decade. Similarly, while Pakistan remained an 'albatross' around India's diplomatic neck, her Afro-Asian allies were reluctant to criticize Republican American administrations for 'tilting' towards Pakistan by rearming her against India. The non-aligned group whether as an Afro-Asian block, the Group of 77, the developing countries, or the non-aligned movement have failed to defend any of their members under attack from the Super Powers — whether it was Nixon's 'sideshow' against Cambodia, or the Kremlin's Kabul misadventure — nor did it succeed in bringing to a halt conflicts between fraternal Islamic States as during the Iran-Iraq or Gulf Wars.

Non-alignment has been ritualized to a point to which it means all things to all States and lacks doctrinal codes of conduct towards the 'faithful' States. Such vacuous posturing might begin to deflect from the real objectives which a State seeks to promote in furthering the security and well-being of its populace. Ironically, India, from being the initiator of this policy has often been its victim, stimulating a readjustment of foreign policy to its domestic economic reforms.

The Ideological World-View

India's foreign policy of non-alignment owed much to an ideological view of the world in which there was not only a bipolar system based on a balance of terror between two nuclear super powers, but also a confrontation of ideologies in which were at stake individual liberties and community rights, cultural regeneration vs social reform, and private vs State enterprise. It is natural that at the time of independence Menon and Nehru would adopt this pre-eminently ideological world-view which arose from their common political environment, if not socialization in a post-Victorian self-righteous and overtly moralistic view of international events. This view has its basis in the British Left's (predominantly Scottish and Welsh rather than English) sympathy with anti-colonialism and the humanism of European socialists. Indeed, the makers of Indian foreign policy were influenced by Bernard Shaw and Leonard Wolfe, the Webbs and Harold Laski, to name a few, as well as lesser-known activists working with the India League, the London Majlis, the League against Imperialism, the Movement for Colonial Freedom, and other Indian nationalist forums in London. In contrast to Menon and Nehru those regarded as having a more pragmatic view, though relatively less publicized as such, were Vallabhai Patel and Syama Prasad Mookerjee. Their geopolitical view of a newly sovereign State had little to do with Curzon's imperial view of India's security and Britain's colonial interests, and here perhaps more in tune with the anti-imperialism of an M. N. Roy or even Palme Dutt. Yet there was an echo from the past policy which could have served India well in Tibet and in security negotiations with China to defining her priorities in rationalistic terms instead of moral self-righteousness.

On the basis of these doubts and concerns, a case can be made for a review of India's foreign policy that does not abandon the hallmark of Menon and Nehru but complements it with the heritage of Patel, Syama Prasad Mookerjee, as well as the universalism of Aurobindo and the realist-power school of Netaji Subhas Bose. Such a review will update and help India to deal with the emerging world order and its main features: the disintegration of India's closest and strongest ally — the Soviet Union; the phenomenal growth of historical cultural-religious allies, Japan and the countries of SE Asia; the need to respond to Europe's renewed interest in India as a fountainhead of civilization and the economic and technological power of tomorrow; acknowledgement of the immense power of the American State, and emphasis on compatibility with its people and institutions confirmed by Indian professionals and students, and immigration to that continent; compatibility which Prime Minister Rao urged during his 1994 visit to the US; and last but not least, to lead South Asia into a new co-operative union that will be focused on rapid economic growth and not further political Balkanization or bureaucratic posturing.

CONCLUDING REMARKS

Culture

There has been an assumption, particularly among Western scholars, that traditional cultures impede modernization policies and are incompatible with science and technology. Implicit in this assumption is that Asian religions are an obstacle to progress and the promotion of human dignity: 'No path led from the magical religiosity of the non-intellectual classes of Asia to a rational methodical control of life'.[5] Such a view has since been discounted by scholars and policy-makers especially in the context of M.N. Srinivasan's views that 'Sanskritization and Westernization are linked processes', and 'the capacity of Hinduism to adapt to changing conditions that casts serious doubt on the widespread belief that Hindu beliefs and practices are a major obstacle to modernization'.[6]

The steady expansion of this view both in India and abroad and the impact of religious revivalism both in Islam and Christianity has altered the nature of the debate. Today, scholarly attention has shifted from the economics to the politics of religion as an obstacle or motive force for national development. Indeed, this question has long been present and the wisdom, or lack of it, in politicizing religion raises the troubling question often fraught with violence: who is being communal and why? Recent world trends about religious revivalism and militancy have thrust this question to the forefront of national politics of secularism, pushing political parties

5 Max Weber: *The Sociology of Religion*, 1922

6 Milton Singer: ed. Myron Weiner, *In Modernization*, 1966

on different sides and often escalating into violence. The resolution to this problem must begin with 'the will to understand in a relaxed manner the inner logic of the different religious traditions [without which] how can one grasp the dynamic of their mutual relations?' (Satish Saberwal, 1986.)

Economics

Globalization has changed the relationship between sovereign States and world markets. Today transnational corporations are seeking to combine national economies into large trading zones which facilitate greater mobility of capital and flexibility of production leading to the accumulation of larger profit margins. One way of achieving this is by making production systems competitive and flexible. The new economic integration of Europe from 1992, the US-led North American Free Trade Agreement, and the *de facto* Japanese-revised co-prosperity zone in Asia exemplify these new trends. While larger trading and industrial blocks are rapidly reshaping the industrial economies of national States, the phenomenal growth of international finance and its international, rather than intra-regional mobility, has lubricated the process of globalization shaking the foundations of old nation States and new regional systems.

Facing this globalization are national economies of 'erstwhile' sovereign States which desperately and hastily seek to change institutions and regulations. In advanced capitalist States it has destroyed neo-Keynesian theories of full employment and threaten carefully-built social programmes of health insurance, workers' employment insurance, minimum and equal wages for men and women, and State and private pensions. In these countries the pressures are not limited to policies of social transformation: pre-industrial free primary and secondary education, food subsidies, preventive health measures, State and co-operative housing, drinking water and power generation. In the name of structural adjustment new policies are also aimed at dismantling collective bargaining which in the West has over the past half century, through its growing linkage to productivity rather than the free wage market, increased the surplus income to the worker and transformed him into the relatively affluent consumer.[7]

7 Drache and Gettler, *States and Markets,* 1992

In this vortex of change, the State cannot blindly shape its policies either according to traditional liberal economic theories or classical socialist economic dogma. It faces unprecedented dangers as well as unique opportunities.

If the growth of international capitalism occurred between the seventeenth and twentieth centuries it was not just coincidental with the expansion of colonialism but as a function of imperialism. Today the new technological prowess of Western multinationals of finance, capital, and management, continually but not always successfully challenged by the new ethics of ecology, redistribution of wealth and alternative value systems, provides a new phase in global relations. It obscures the sovereign political and military power of weaker States achieved in the course of anti-colonial struggles.

It is in the global context, of what is currently unfolding rather than expectations and goals of the past, that the wisdom and capabilities of India's political forces will be tested in the efforts to shape the new strategies for productive efficiency and social advances. In this crucial task of reconstituting India it is incumbent that the leadership constantly review its popular support while resolving threats to national unity.

I would especially like to acknowledge the following: two friends from undergraduate years at Calcutta University and London University, Naranarayan Gooptu, Advocate-General of West Bengal for shared experiences in the student movement and a 'world view', and Amal Datta, Member of Parliament, for shared economic insights and a 'global technotronic view'; dear friends in Canada, the late Dr Madan Handa, Professor, OISE, Toronto and the late Mrs Barsha Mitra-Kelly, Executive Secretary, Canadian Association for South Asian Studies; M. Vinayak who has been keeping me in touch with political and business leaders; Eleanor Maunula and Karen Woychyshyn for valued secretarial help; and Shyamalendu Banerji for helping with revisions.

The State in Transition

RAJNI KOTHARI

Of all the uncertainties facing India, perhaps the most crucial one is with respect to the condition and role of the State. This is partly because the first-generation leadership had decided to entrust the State and its institutions with the task of building a nation that was to be knit together as one entity in and through its great diversities, in the process becoming democratic, just, free from vested interests, respectful of diverse traditions (which is what we had meant by the term secular), and one in which all people, irrespective of their other differences, would be able to actively associate themselves. If today the State in India is unable to live up to this mandate, as it has appeared to over the last several years, there is reason to be concerned; and, if indeed it were to move away from that mandate, to be clear about why this was being done and, depending upon the choice we wish to make at this juncture in world history, accept or reject such deviation from the original premises. It is as yet not clear whether India will move along a different course than what it had earlier set out for. There are major pressures at work to take it along a different course, but, as of now, the State in India is still in a state of tremendous uncertainty as regards its role in shaping the present and the future of this country. Before going into the condition one finds it in at the present juncture, it would be useful to briefly understand the original conception.

HISTORICAL OVERVIEW

Ever since Independence the State has been conceived as a Nation State. On the one hand this involved the task of building a single, new, 'nation', encompassing several identities and affiliations: the relationship of the new nation to pre-existing identities was to be one of transcendence though it was also one of encompassing them and in some ways even 'representing' them all in a composite manner. On the other hand this very conception of encompassing diverse identities — social and regional, linguistic and cultural — meant assigning a key role to the State through which the composite nation was to be created. In effect it was 'Stateness' that gave to the new entity an encompassing, representative and transcendent quality. It almost reminds one of the old Hegelian idea of the State although the authoritarian overtones of such an integrationist ideology came much later as did the assertion of various pluralities which together produced a crisis of representativeness of the Indian State.

For a long time the nation-building exercise was conceived to be one that would lead to greater unity of the nation and greater involvement and participation for the citizen in the institutions of the State — hence the conception of an 'open society' and the belief that democratic participation was the best way of achieving national integration. The elite undertook to promote democracy through the very composition and institutional structures of the State: the Parliament and State legislatures; cabinet governments at different levels which were to promote 'consensus' through representing diverse regions and interests; an administrative structure that was to be accountable; a federal continuum through which such governments and administrations were to operate; and a judicial system that was to ensure that all these institutions operated responsibly and according to the 'rule of law'. There was also a unique party system which saw to it that the relationship between the Government and the social structure was a continuous one, both vertically and laterally, and was mediated by a number of intermediate institutions and elites: it was through the party system that the colonial 'steel-frame' was conditioned by political processes and actors.

The elite itself chose to relate to diverse interests in society as well as to its traditional voluntaristic infrastructure, thereby narrowing the distance between Government and the people and reducing mutual alienation that was inherent in the imported Westminster model. Inherent in such a conception were two seemingly opposite ideas but which were meant to serve the common purpose of building a democratic Nation State. The first of these was to assign primacy to the State in the whole process of nation-building. Inherent in the first, was a decentralized conception of sharing power and carrying diverse entities in a common enterprise. The second, on the other hand, left to itself, could produce a centralized model of governance. But the two together were meant to produce a workable balance between a strong and coherent Centre and a series of vibrant sub-centres representing the large variety of structures and spaces through which a composite Nation State was to take effect. The nationalist ideology that had accompanied such a conception was both integrative and open to diverse interests and groupings, including dissenting ones.

This model of a Nation State is today in disarray. On the one hand, the State is ceasing to be the central mechanism of the nation, while on the other, the 'nation' that had come into being has entered a process of disintegration. Both the retreat of the State and the disintegration of the nation are going hand in hand: one being hijacked by the global 'order' and 'market'; the other facing serious challenges, within, from the large number of peripheries, as well as externally, from different regions and 'nationalities'. With this the project of nation-building and, as part thereof, the central role of the State in social change is likely to be given up. As this happens, the autonomy of civil society too will decline and with it will also go down the modern search for a democratic order.

Alongside such globalization, there has emerged what may in theory appear to be its antithesis: the rise of ultra-nationalism, that, like globalization, seeks to suppress various pluralities though more along a religious direction than an economic or technological one. It will also pitch Nation States against each other and internally provoke a series of sub-nationalisms representing various suppressed or latent nationalities and ethnicities.

It appears now that the early modernizers who thought of creating a modern Nation State in India (as in many other countries in the Third World) had an inadequate understanding of the contradictions that were inherent in such an enterprise, particularly given the fact that the country was full of not just diversities but also disparities which, far from being reduced by the new Nation State, have been accentuated. As the country moves towards the twenty-first century, it is going to face this shift from integrative to disintegrative tendencies more sharply than was the case during a large part of the twentieth century.[1]

THREE 'CRISES'

We can now examine the different facets of the 'crises' in which the Nation State in India finds itself today.

First, there is a crisis of legitimacy, both in terms of the regard and respect for the State in the eyes of the people of India, and in its credibility in the eyes of others. Second, there has been over the years, a gradual breakdown of authority, vis-à-vis citizens at large, and of the Centre vis-à-vis intermediate and local institutions — the States, the local centres of power, community organizations, voluntary structures and other institutions of civil society. Third, there has been taking place an erosion of the autonomy of the Indian State vis-à-vis other centres of power within the country (industry, labour, religion, caste, linguistic identities and other interests) and external interests, which has led to a declining emphasis on things like self-reliance, non-alignment, and playing a leading role in the Third World.

RETREAT

From this, it can be observed that the State in India, instead of occupying centre stage in the social and economic arenas, is seen to be in retreat. This is obvious in the economic sphere from which it has been regressing for over more than a decade now and, of course, most dramatically during 1991-2. It is also true in the social sphere in matters like providing protection to the minorities, the backward classes, tribal communities and women and children; encouraging people's initiatives and indigenous thinking in key areas like education, environment, health and housing; and generally implementing the laws of the land for the betterment of the poorer and oppressed sections of society.

1 I have argued elsewhere – *Seminar*, January 1992 – that the twentieth century has already come to an end.

However, the consequences of this regression are subjecting the State to increasing challenges from both ethnic and regional movements and class and caste arenas in the struggles for their democratic rights, leading to a growing sense of disintegration of the polity.

Yet another consequence of the retreat of the State has been its growing decline as an engine of development, leading to a sharp undermining of institutions meant to promote planned development with a view to fulfil social obligations as enshrined in the Constitution of India: in its Directive Principles, under Fundamental Rights, as well as under many other articles of the Constitution.

'INTEGRATION' INTO A 'WORLD ORDER'

Pari passu with this retreat of the State domestically has been its decline as a political entity internationally. Here one finds that far from continuing to assert its identity and independence vis-à-vis dominant centres of power in co-operation with other countries of the South, the Indian State is being prepared by its bureaucratic, industrial and technocratic elite to become integrated into a 'world order' that is being designed under the tutelage of the United States. Though it has so far been presented in terms of integration into the world market, still assuming that economic integration need not lead to political dependence or social and cultural absorption, an assumption that seems to me to be at once naive and deceptive, the general consequence of such integration is a clear loss of autonomy of the Indian State. Incidentally, in this regard, the retreat of the State and the centrality of the market need not be conceived as opposed to each other. What happens in reality, as found for instance in South Korea, Taiwan, Japan and more recently in Malaysia, is that the State and the market reinforce each other under the domination of an *elite* that succeeds in undermining both dissent in the political sense, and institutions of civil society in the social and cultural sense.

Those who believe that terms like liberalization and privatization simply mean replacement of the institutions of the State by those of the market seem to be living in a make-believe world that has been promoted by the information and communications blitz let loose by global centres of corporate power. The fact that people begin to accept such make-belief is understandable, given the vacuum — both ideological and political — that has followed the virtual dissolution of the Third World (both as an idea and in reality), the actual (not just virtual) dissolution of the socialist block and, in quite a different way, the decline of the Nation State in Western Europe following its integrationist ideology. The philosophy of liberalization and privatization comes in handy in the face of such a vacuum, just as the decline of real democracy in plural societies like ours has created a vacuum that is being filled in by religious and sectarian ideologues.

FUTURE PROSPECTS

It is with respect to this decline of the Nation State as an instrument of social transformation on the one hand, and as an embodiment of distinct (if not 'sovereign') political entities on the other — and this in the larger context of a growing ideological vacuum globally — that one has to assess the prospects for the future. There is the likelihood of moving away from the conception of a world composed of distinct nation States and diverse political cultures towards one that calls for their 'integration' into a so-called world order. On the other hand the same Nation State will be subjected to increasing challenges from the grassroots, and diverse peripheries and regions, both these tendencies eroding the autonomy and integrity of Nation States like India. The integrative trend is going to be backed by organized interests and resources, both financial and political, while the democratic trend towards a more pluralist and decentralized structure is likely to be backed by the power of numbers and their growing consciousness of rights. The first (integrative) trend will be for creating a common future for the global middle class and will end up leaving a large part of humanity out of the historical process, out of both the institutions of the State and the transactions of the market. It is also a trend that is likely to produce an increasingly centralizing, homogenizing and a potentially authoritarian structure of governance. Also, the more the integration of the Nation State and national economy with the global system, the more the alienation of the ruling strata from the peoples, communities and nationalities round the world.

The same is true for within the country: the more centralized the State becomes, the more distant will it get from the grassroots and the diverse communities and nationalities. As both these trends (global, and within the country) converge, large sections of the people will be left with no option but to join a growing culture of revolt and rebellion, strikes, *dharnas,* and *bandhs*, some of them peaceful, but a growing number adopting the path of violence. The argument, that those at higher levels tend to be more enlightened than those at local levels, that the proposed new structures (like panchayats) will be dominated by dominant social groups (castes), may be valid for some time but not for long: once the democratic process gets going, the more numerous social classes will feel empowered and challenge the hegemony of the upper strata. The same is the case with the argument about global institutions and elites being more enlightened than national and regional ones – hence the need to transcend the Nation States and other 'parochial' interests and institutions.

This argument was advanced in all earnestness after the First World War mainly with a view to restrain the aggressive element in the human being and create a more peaceful and progressive world. Today there are not many takers for this argument which has been rejected not only for being too romantic, but also because the underlying world-view has been hijacked by global capitalism and its effort to colonize the whole world once again. It is also a world-view that is being advanced to contain the democratic upsurge round the world through an ideological package (liberalization, privatization etc.) for integrating the middle and the upper classes everywhere with the simple aim of retaining and enhancing the privileges and perquisites of these classes. There should be no mistaking the political motives behind the emerging attempt to defeat the struggles of the Third World States seeking autonomy and self-determination.

There has been a growing debate over a long time about governance and stability wherein several Western intellectuals have argued for retaining a balance between the governed and the governors, who need to retain an edge, so that the forces of revolt and 'chaos' do not undermine the stability of the *status quo.*[2] This was for long an academic debate, carried out by academics. It has now got a lease

2 Samuel Huntington and other Western scholars, worried at the growth of dissent in Western societies and of Third World challenge to Western hegemony, have articulated this position. See their 'Report to the Trilateral Commission' along these lines.

of life with the proponents of an integrated world order and a global market making it appealing to elites of the Third World as they face growing restlessness and turmoil from the grassroots of their diverse societies: India has of late joined this bandwagon of integrating with the world market and becoming part of a new world order.

EMERGING TRENDS

It is in light of this growing tendency of integrating the whole world in the framework of world capitalism and Western hegemony that we have to think of the likely future of the Nation State in India and in similarly placed countries of the Third World. Three important trends need to be taken note of. First, there is a marked rise of subnational consciousness overtaking the earlier kind of nationalism that included diverse regional and cultural identities. There is a growing revolt against central authority and, in some areas, against State Governments, producing a highly unstable situation. The 'Centre' is increasingly unable to hold against the growing challenges to its authority, its nervous attempts to suppress this by recourse to paramilitary forces and the Army (whose increasing use in one region after another is also proving an unsettling factor) only underlining its inability to hold.[3]

Second, the national elite, as it becomes more and more unable to cope, tends to turn away from the basic political challenge of nation-building and the related economic challenge of 'development' by intervening against vested interests, reversing the large disparities that exist and are growing, and undertaking a model of growth that fulfills the minimum needs of the people and reduces inequities and injustices: in the process taking along dissatisfied social strata and regions, and minimizing social tension and violence. The new generation of elites seems to be giving up on these two challenges — nation-building and development — and turning to international and transnational agencies for bailing them out. The growing accent on technology as the pathway of development has necessarily led to moving away from self-reliance and towards dependence on external forces (which is then called 'interdependence'). It has also led to moving away from political representatives of the people to handing back to the bureaucratic *apparatchiks* of

3 See my 'The Centre Cannot Hold', *Seminar*, January, 1992.

post-colonial State, very much in line with the colonial legacy. It has in fact become doubly colonized by deepening the dependence on external forces, through local elites, willingly submitting to their dictates in deciding our priorities and objectives.

Third, there is a rise of new organizational forms that are transcending States and nations, representing almost a mutation in human evolution. These consist of transnational corporations possessing enormous financial and technological resources and a series of bilateral and multilateral compacts providing for regimes of security aimed against insurgent movements from below, together producing a new institutional apparatus of running the world through a so-called 'world order'.

CHALLENGES FACING THE INDIAN STATE

It is against these changes in the historical context of a Nation State like India, especially given its multicultural and 'multinational' nature, that its survival will depend. To a large extent the future of the Nation State in India depends on whether it becomes still more centralized and run from the 'Centre' — which then makes it an alien presence and also makes it an easy target for global interests — or becomes increasingly democratic internally, and on the basis of that inner strength plays a democratic role externally. The Nation State, a highly complex world full of contradictions and tensions, has to become a symbol of voices of diverse cultures and peoples, and thus give rise to a dispersed and decentralized world order, preserving the integrity and the uniqueness of each, pursuing distinctive goals and ideologies, giving rise to a consensus that is derived from respect for such diversities. Its role should be to intervene in the larger world and the various global forces at work. The challenge it faces is to restore this historic role from which it seems to have strayed away for the past several years.

The challenge to restore such a role to the Nation State faces not just national and state level elites but also the various sub-national identities and movements. There is need for the latter to realize the real sources of exploitation and homogenization of all nations and peoples which happen to be increasingly global. It is in the creation of a multicultural and plural State organized on federal lines that their goal of autonomy and self-determination can be better realized than by undermining the legitimacy and authority of the State. Fighting against the State as such will only reinforce the latter, opening its doors to transnational and global enterprises and military-security compacts. Once this process gets under way (it has only begun of late), there will be little scope for the success of democratic movements within the country. Undermining the legitimacy of the Nation State will only facilitate its surrender to global actors who are yearning to make the whole world accept their model of 'world order'. It is necessary that the increasingly beleaguered States, the increasingly oppressed social strata, and ethnic and regional groups, work towards a common future which will enable them to withstand the challenge from transnational and global actors.

But, of course, for this to happen, for the movements to trust the Centre or even the State centres, the Nation State needs to be organized differently than has been the case so far. The basic task facing the national elites is one of regaining legitimacy for the State, accepting for this purpose the reality of it being a plural entity, restructuring internal democracy along federal (if need be confederal) lines, and resisting external pressures for downgrading the State in the development process. In short, there is need to think of Nation States like India to be elements in a large federal continuum: from local (village to district), to sub-regional (states, sub-states), to national, to supra-regional, to global. Such a federal conception will also involve a federation of diverse social strata many of which are still suffering from grave injustices and growing disparities. Incidentally, this is also the only way of defeating fundamentalist and fascist tendencies based on religious and other sectarian appeals.

In moving towards such a structure of the State, as well as the world at large, it is necessary not only to restore to the Nation State its historic role based on the clear acceptance of the federal perspective on democracy, it is necessary not to be taken in by the growing charisma of the market — the ideology of privatization, the miasma of consumerism, and the yuppie culture. In India the power of the market forces and consumerism can already be seen even in the State-run Doordarshan. Many other institutions of planning, development and finance are also already being run at the behest of the corporate sector. The power of the private sector is already the upswing, alongside the growing role of the mass media controlled by it and the ad agencies through which people's choices and tastes are being shaped. Given a chance (now especially with the emphasis on privatization), these forces will replace democratic politics by emphasizing technological fixes.

It is important for us to remember that even in the West it became necessary for the State on the one hand and trade union and consumer — and lately environment and women's movements — on the other to ensure distributive justice, environment ethics and affirmative commitments obligating industries, professional business houses, universities, science and technology establishments and market-related agencies to abide by the laws of the State. We must also not forget that market forces in the West had let loose oligopolies and cartelized speculative share markets and trade cycles that produced the Great Depression from which Keynesianism proved to be the only way out. It is important for us to understand the pathology of a development process based on the market before we decide to throw away the advantages of a comprehensive process of nation-building and a corresponding political economy in which a central role is assigned to the State. We must, of course, deal with the pathologies released by the State but not fall for still worse pathologies.[4]

It is in the manner in which India's political elite, party system, and people's movements respond to these challenges — of both thought and action — that the future of India as a Nation State will depend.

4 On this whole issue, see my 'State and Statelessness in Our Time', *Economic and Political Weekly,* Annual Number, vol. XXVI, nos. 11, 12, March, 1991.

Indian Polity Today

MADHU LIMAYE

The transfer of power from British to Indian hands on 15 August 1947 was by and large peaceful, although it was accompanied by violence and unprecedented migrations from and to the north-western and, on much lesser scale, the north-eastern parts of India.

The Government of India Act 1935, as modified by the Indian Independence Act of 1947, served as the framework of the new State pending the adoption of a new Constituent Assembly. The Constituent Assembly could not ignore the constitutional developments in the preceding hundred and fourteen years. In order to understand the evolution of the Indian polity after Independence, a brief review of the previous constitutional developments would be quite enlightening.

When the British established their control over the greater part of India, a provincial system gradually grew up. But this provincial system was not organized on any rational principle. The map of the provincial system in British India was the product of military exigencies and administrative convenience. The Act of 1833 secured the subordination of the three Presidencies, Calcutta, Bombay and Madras, to the Central Government. The Directors wanted the Governor-General to consider to what extent the powers of the Government could be best exercised by the local authorities and to what extent and in what particulars they were likely to be best exercised and retained in the Governor-General's hands. They suggested at the same time that interference in the minute details of provincial administration by the Governor-General would be neither desirable nor practical.

In the early period of the British Raj the three Presidencies had been endowed with the power to make laws independently of each other. The Presidencies of Bombay and Madras sent copies of the acts, rules and orders passed by them, but the Government of Bengal did not possess the power to modify such legislation. The Act of 1833 deprived the Presidencies of legislative power — all they could do was to make proposals for legislation — and it made the Governor-General's Council the one supreme power competent to enact laws for the whole of British India. However, the Council Act of 1861 introduced two changes: one, it provided for the addition of a 'native element' to the Councils; and two, it restored to the two Presidencies the power of legislation which had been taken away twenty-eight years earlier, but was to be subject to the assent of the Governor-General. This system prevailed till 1919.

The dominant idea during the years of British rule in India (right up to 1919) was that the entire governmental structure was an indivisible whole and, in the final analysis, subject to Parliamentary control in Britain.

The Decentralization Commission recommended that future policy should be directed to steadily enlarging the spheres of the detailed administration entrusted to the Provincial Governments and the authorities subordinate to them and that these governments and authorities should dispose of an increasing share of the ordinary work of administration.

An innovation of far-reaching significance, similar to the addition of an Indian element in the Legislative Councils, deserves mention: it was the gradual ousting of Muslim and Hindu Criminal Law and its substitution by the British system of jurisprudence. The rule of law, the establishment of the High Courts, and introduction of prorogative writs, including the famous writ of *habeas corpus*, together with great enactments like the Criminal Procedure Code, the Indian Penal Code and the Evidence Act based on the principle of equality before the law, made a profound impact on constitutional evolution in India These concepts and rights came to be incorporated in the new Constitution.

The emergence of the Indian National Movement, with the Congress as its supreme embodiment, in 1885 accelerated the process of change in several directions. The more insistent nationalist demand was for the progressive Indianization of the services, and admission of Indians into the higher echelons of the administration, including the exclusive Indian Civil Service. Another demand was for the strengthening of representative institutions at all levels, introduction of the elective principles, and conferring of greater powers on these institutions. Yet another stand in the nation's demand was devolution of power and a substantial Indian share in the administration.

The leadership of the national movement was not, however, clear about the form of government that they wished to see established in India. A memorandum submitted by the non-official members of the Imperial Legislative Council towards the close of 1916 — among the signatories were Tej Bahadur Sapru, Srinivasa Sastri, M.A. Jinnah and Nazrul Haq — demanded that all the Legislative Councils in India, at the Centre, and in the provinces, should have a substantial majority of elected representatives. The memorandum wanted the franchise to be broadened. The document also said that half the members of the Executive Councils should be Indian, and, further, that the elected representatives of the people should have a voice in the selection of the Indian members of the Executive Councils and that for achieving this end the principle of election should be adopted.

The Congress-Muslim League scheme of 1916 (adopted in the Lucknow session of the two organizations) went into the question of Muslim representation in the various Provincial Councils and the Central Assembly, and reiterated the demand voiced in the aforesaid memorandum for a substantial elected majority in the Legislative Councils. It desired that the Indian posts in the executive Councils at both Central and Provincial levels be filled by the elected representatives through an election. The Congress and the League asked that the popular element should constitute half the total strength of the Executive Councils.

From the foregoing it would seem that the Indian leaders were not looking forward, at this stage, to the establishment of a parliamentary form of government. They had in mind something close to the Swiss model in which the Executive is filled by election. The next instalment of reform, after the working of the Congress-League scheme for a reasonable period of time would, the Indian leaders hoped, be complete Indianization of both the Legislative and Executive Councils.

In response to the agitation in India for Home Rule, the Secretary of State for India announced on 20 August 1917 that the policy of the British Government pertained to an increasing association of Indians in every branch of administration and gradual development of self-governing institutions with a view to bringing about a *progressive realization of responsible government in India.* The Montford Report described this declaration as the most momentous utterance ever made in India's chequered history.

Making some allowance for exaggeration, it was quite true that this declaration, and the Montford Report together with the Government of India Act based on it, constituted a turning point in India's constitutional development.

Although the Congress under Mahatma Gandhi turned to non-co-operation and refused to put through the reform scheme — known as dyarchy — a section of the Congress constituted themselves into a Party and decided in 1923 to end the boycott of elections Thus emerged the Swaraj Party.

Dissatisfaction with the tardy pace of reform mounted in India. The Government of India Act 1935 sought to reorganize the Indian polity on the basis of the federal idea, with an elaborate and exhaustive distribution of legislative powers between the federal centre and the largely autonomous provinces, and a federal court to adjudicate constitutional disputes. The statute abolished dyarchy in the provinces and introduced the principle of responsible Government (barring the undefined area of the governor's special responsibility) in the provinces. The franchise was also broadened, although adult franchise was still a far cry.

The Congress decided to contest the elections to the new Provincial Assemblies to be held in early 1937, and emerged victorious in all the provinces.

The Congress system of organizational control was so effective that there was no possibility of governors playing politics or trying to bend the administration to their will. The Congress Legislature parties were disciplined bodies and the ministries functioned on the basis of collective responsibility.

The War broke out in September 1939 and the federal part of the 1935 scheme was indefinitely postponed.

Congress leadership was by now fully committed to the parliamentary form of government. It rejected Cripps's offer in 1942 largely because the new government which was to have been set up at the Centre would not function as a Cabinet in a normal constitutional government and would continue to be hindered by the Viceroy enjoying all his old powers. When the final phase of constitutional negotiations opened in 1946 and an Interim Government was formed, the Viceregal Executive Council, with Jawaharlal Nehru as vice-president, tried to function as a Cabinet on the basis of collective responsibility.

The exclusion of the Muslim League from the ministries, under the 1935 Act, in the Congress provinces and the fear of the vast and growing power of the Congress after 1937 gave momentum to the Muslim League agitation and by the end of the War the League established its supremacy in the Muslim majority provinces (minus the NWFP). In the end it insisted on the partition of India.

Jawaharlal Nehru and Sardar Patel, in a bid to stop the fratricidal war that disfigured the north-western and eastern parts of India, finally decided to accept the division of India. Power was finally transferred to the Indians and two Dominions came into existence on 15 August 1947. The Indian Independence Act made the Constituent Assembly a sovereign legislature; the head of the Government was now called Prime Minister and other ministers were appointed on his advice The special powers and independent judgement and discretion exercizable by the Governor-General and Governors were abolished.

India was to be a federal polity in which the 'Executive should be of the Parliamentary type'. The suggestion by some Muslim members that ministers should be elected was rejected.

It logically followed from this decision that the President should function as a constitutional head and exercise his functions on the advice of the ministers. He was not to have any discretionary powers nor was he to be elected through a direct election based on adult franchise. An electoral college, consisting of the Members of Parliament and State Assemblies, was to elect the President for a five-year term.

The Constitution cast on the Prime Minister the duty to communicate to the President all Cabinet decisions and information relating to the affairs of the Union and proposals for legislation. He could also require his Council of Ministers to consider decisions taken by a minister which had not been cleared by the Council. This was an important power and enabled a President, with the necessary stature, to exercise a restraining influence over the Cabinet Ministers. The Constitution expected the Prime Minister to extend to the President every courtesy, and despite differences of temperament and approach, successive Prime Ministers, with the sole exception of Rajiv Gandhi in 1985-7, have managed their relations with Rashtrapati Bhavan rather well, without allowing their differences to surface.

When the question of the mode of the formation of a Council of Ministers in the provinces (called the States under the new Constitution) came up for consideration some Muslim members of the Constituent Assembly tried to revive the idea of ministers' election by the State Legislature on the basis of proportional representation (PR) to secure effective minority representation. But Sardar Vallabbhai Patel shot down the proposal. The Governors too were to be appointed by the President — not elected — and hold office during his pleasure.

In subsequent years the question of the status and powers of the President — and Governors — became a subject matter of debate. On 21 March 1950 in a letter to Prime Minister Nehru, Rajendra Prasad, the first President of the Indian Republic, questioned the doctrine that the President's position was similar to that of the Crown under the British Constitution. The Attorney-General, Motilal Setalvad, and Alladi Krishnaswamy Ayyar, an eminent jurist, gave the opinion that the President could not act independently; that he was a constitutional head and under Article 74 was required to exercise *all* his functions on the advice of his ministers. They reinforced their argument by citing the corresponding provision of Article 163 which had made a significant distinction between the Governor acting within his discretion and his acting with the aid and advice of his ministers. There was no such distinction at the Centre.

Frustrated politicians, elitist groups excluded from the new power structure, and opposition groups repeatedly defeated in elections, however, continued to raise the question of Presidential powers, including his power to dismiss the Prime Minister. This issue did not assume much significance during the Nehru era. But Nehru's daughter never felt as secure as her father. It was the fear of a Presidential *coup d'état* engineered by the dominant Congress group — the Syndicate — that prompted her to defeat her own party's official Presidential candidate in 1969 and secure the election of V.V. Giri who she thought would be a docile President. This led to a split within the Congress. In the 1971 mid-term Lok Sabha election Indira Gandhi won a convincing majority.

But she still continued to be haunted by the threat of Presidential interference. During the Emergency she introduced drastic constitutional changes and sought, through an amendment of Article 74 dealing with the Cabinet (42nd Constitution Amendment Act) to place beyond all doubt the binding character of the advice of the Cabinet. Morarji Desai, Prime Minister in the Janata Party Government, was also not entirely free from apprehensions about Presidential manipulation and so he refused to restore the Article to its old form.

A significant constitutional development in the post-Nehru era was the gradual extension of the Prime Minister's power exercised through a powerful Prime Minister's Secretariat or Personal Office (PMO). Jawaharlal Nehru had himself planned to create such a Secretariat but Lord Mountbatten had persuaded him to drop the idea. His successor, Lal Bahadur Shastri, lacked Nehru's confidence and therefore created the Prime Minister's Secretariat to strengthen his hold over the administration. Indira Gandhi can legitimately be said to have perfected this instrument of control, but the subsequent Prime Ministers also kept the PMO system in being. The PMO soon started passing instructions to the departmental officers, and the ministers were often bypassed.

It was felt that the co-existence of a Governor elected by the people, and a Chief Minister responsible to the elected legislature might occasion friction. It was therefore decided to make the Governor's office appointive. The Governor, it was laid down, would hold office according to the pleasure of the President.

However, nobody then imagined that this provision would be wilfully abused and that the Governor's office would be made part of the patronage pool and then of the spoils system. Nehru laid down certain norms for appointment to this office.

In practice, Nehru himself departed from this norm and the office steadily became part of the patronage pool. Defeated politicians, favourite retired civil servants, diplomats and generals came to be appointed Governors. Although the Supreme Court held that Governors were constitutional heads of the States and in no way subordinate to the Central Government, the Congress Party always wanted the Governors to be pliant instruments of their policy. Indira Gandhi made an example of two Morarji Desai appointees in 1980 – Raghukul Tilak and Prabhudas Patwari – by dismissing them summarily. The National Front Government headed by V.P. Singh emulated her and went in for wholesale dismissal of Governors. It acted on the American principle, 'The Victor takes all'! It was thought that P. V. Narasimha Rao's Government would make a healthy departure from this unsavoury practice. But the hope has been belied.

There have been important departures from the course of constitutional developments in the British days. The first important change was the abolition of special privileges for religious minorities – Muslims, Sikhs and Christians. However, the Scheduled Castes and Scheduled Tribes among the Hindus and Sikhs received powerful constitutional protection. There would be reserved seats for them in the elected legislative bodies in proportion to their population and they would also have representation in the services. In addition there were to be special provisions for the 'socially and educationally backward class of citizens' (meaning the backward castes among the Hindus and lowly occupational groups among the Muslims and Christians). The administrative services and the judiciary were also important elements in the structure of power. And so it was necessary that the disadvantaged sections secured adequate representation in the administration at all levels of the edifice of power.

Abolition of untouchability and emancipation of women were important planks of the national movement. The Communist ideology accepted the primacy of economic interests and laid emphasis on class struggle. The Marxists never addressed themselves seriously to the problem of social inequality. Nehru also tended to stress the importance of the economic factor and did not have a genuine understanding of the suffering of the socially disadvantaged groups as distinct from the problem of poverty. Among the Socialists, only Rammanohar Lohia tried to understand this problem and made the doctrine of special opportunities an important part of the programme of social reconstruction. He tried to integrate the Phule-Ambedkar heritage with the Socialist Movement. The Congress Party early became an effective vehicle of the aspirations of the backward classes in the southern States and acquired a formidable mass base in Andhra Pradesh and Karnataka. However, they did not implement the reservation policy in the Hindi-speaking States or at the Centre. Sometime in the mid-fifties Nehru turned his back on the reservation policy, and the backward classes began to desert the Congress in north India and look to the SSP, the Janata Party, the Lok Dal and the Janata Dal to realize their aspirations. The attempt of the Janata Party and the Janata Dal to introduce reservation in Bihar and UP in 1978 and at the Centre in 1990 respectively provoked violent reaction. The problem persists even today.

The Nehru Government undertook in 1954-6 work of social engineering of far-reaching significance in the areas of marriage, divorce, adoption and inheritance. The first to be put on the statute book was an updated version of the Special Marriage Act which permitted intercommunity marriage without forcing the parties to renounce their religious faith or compelling either spouse to embrace conversion to another religious faith. The new Special Marriage Act (plus the Indian Succession Act which has been on the statute book for many years) constituted a voluntary Common Civil Code. The attempt of Dr Ambedkar to secure the passage of a comprehensive Hindu Code Bill in 1951 was frustrated by the opposition of President Rajendra Prasad, backed by a large number of conservative members of the provisional parliament. Prime Minister Nehru beat a retreat in the face of this opposition but in the new Lok Sabha, elected on adult suffrage, he pushed through the project of codification of Hindu Law in instalments. Before the election in 1957 the new structure of Hindu Law was in place Nehru did not follow up this great work with the necessary reform of

Muslim and Christian Personal Law. Non-action in this area has been a glaring failure of the Indian polity.

The Communist Party was indifferent to the question of social reform; neither did the Socialist members of Parliament make any valuable contribution to the debate on these enactments. As to the Jana Sangh, it was totally opposed to the codification of Hindu Law and considered it to be against Hindu dharma and culture. The JS-BJP subsequently took up the issue of a Common Civil Code, not out of conviction, but to damn the Congress as a protector of Muslim obscurantism and appeaser of Muslim-vested interests.

Centre-State relations after Independence involved not only the powers of the constituent States, but also the question of the redrawing of the administrative map of India. *The State administrations are the foundation of the Indian political system*. It was through the States that the aspirations of the newly politicized sections of our population could be satisfied. India is a country of unimaginably vast linguistic, cultural, regional, religious and ethnic diversities. The spread of political consciousness among different groups has been uneven. Newer and newer groups and regions are being drawn into the stream of political awakening.

When the Congress became a mass movement, its leaders like Lokamanya Tilak and Mahatma Gandhi naturally had recourse to people's languages as the vehicle of political propaganda. This in turn gave rise to a new consciousness among the various language groups inhabiting the country. The British Indian provinces were not created on the principle of linguistic and sub-cultural homogeneity. The Bombay Presidency embraced four language groups: Sindhi, Gujarati, Marathi and Kannada. The Madras Presidency too was full of diverse language groups, the major groups being Andhras, Tamilians and Malayalis. But the Congress as a mass movement recognized early the need of organizing its work on the basis of linguistic units. The first to demand a separate Provincial Congress Committee was Sind, followed by Andhra. Gandhi went further. When he reorganized the Congress, the PCCs were largely created on a linguistic basis. The reorganization of States on a linguistic basis was a cardinal principle of Congress policy. And yet when the country became independent, Nehru and Patel, who were grappling with the stupendous after-effects of partition, shrank away from tackling the issue of States' reorganization.

Sardar Patel and Nehru accorded the highest priority to the task of integrating Princely India into the Indian Union. Amalgamation of a large number of Princely States into compact new States was no mean achievement of the Indian State. The creation of Rajasthan, Madhya Bharat, Vindhya Pradesh, Saurashtra and Travancore-Cochin constituted significant benchmarks in the consolidation of the Union.

Both Patel and Nehru thought that States reorganized on linguistic principles would breed provincialism, fissiparous tendencies and conflict. Unlike Gandhi, they feared strong provinces (States). In the case of Nehru, the influence of Marxism caused him to emphasize economic interests and class identities to the exclusion of other identities.

The death of Potti Sri Ramalu, the respected leader of Andhra, in the indefinite fast undertaken by him, however triggered a violent movement, and Nehru came to the conclusion that the establishment of a separate Andhra State would serve the interests of India's body politic. He also set up a States Reorganization Commission to examine the whole question.

The vast area from Rajasthan, Punjab, and Jammu and Kashmir in the north and west, to Assam, West Bengal and Orissa in the east was not much affected by the reorganization of States. But in the south there was a drastic rearrangement of the States. Also there was created in the central parts of India a sprawling Hindi-speaking State of Madhya Pradesh. In this were included Madhya Bharat, Vindhya Pradesh, Bhopal and Hindi-speaking parts of the old bilingual Madhya Pradesh State. The old Hyderabad State was broken up and its Telugu and Kannada-speaking districts were joined to the States of Andhra Pradesh and Karnataka. The attempt to set up separate States of Telengana (Telugu) and Vidarbha (Marathi) was given up. The new State of Andhra Pradesh united within itself all the Telugu-speaking areas, and Karnataka now comprised the old Mysore State and the areas detached from the State of Bombay, parts of Madras and the three districts of Hyderabad. Parts of Travancore–Cochin, predominantly Tamil-speaking, went to Tamil Nadu. Malabar was detached from Madras and the new State of Kerala now gathered together all the Malayalam-speaking people in a single State.

Contrary to the fears entertained by Nehru and Patel, this rearrangement of the administrative map strengthened the foundations of the Indian polity. The major dissatisfied area was the Marathi-speaking part of western India. The Marathi-speakers doggedly resisted (a) the creation of a separate Marathi-speaking Vidarbha State instead of joining a united Marathi-speaking unit; (b) the attempt to separate the city of Bombay from Maharashtra; and, finally, (c) the experiment of the greater bilingual Bombay State which brought all the speakers of Gujarati and Marathi together in one huge administrative unit. The agitation for a united Maharashtra continued resulting in electoral losses for the Congress. It was Indira Gandhi who as Congress President persuaded her father and other Congress leaders that resisting the inevitable breakup of the greater bilingual Bombay would be productive of great mischief. In 1960 Maharashtra was finally separated from Gujarat. There now remained only two areas where the demand for rearrangement of States was met with a firm rejection: East Punjab and Assam. Nehru was hostile to the demand of the Akalis for the formation of a Punjabi-speaking State. He thought that the slogan of a Punjabi Suba was really a demand for a Sikh State. But the regional formula (under which the Punjabi-speaking and Hindi-speaking areas were to enjoy some autonomy) did not work, and after Nehru's death in 1964, and the conflict with Pakistan in 1965, there was a visible shift in the Centre's position and the decks were cleared for the formation of Punjab and Haryana. This division proved beneficial from the point of view of economic development. But even the formation of a separate State in which the Sikhs were a majority did not satisfy the Sikh's political aspirations because (a) the Congress Party, for its own partisan ends, brought about the downfall of popularly-elected Akali-led coalitions either through defections (1967) or the device of President's rule (1971, 1980 and 1987); (b) the misconceived notion of Sikh identity being threatened by the Hindu ethos; and (c) Pakistan, and the emigrant Sikhs, ceaselessly fed the insurgent elements from outside.

The Assamese people strongly opposed the demand of various tribal groups for separate States of their own. The provisions of the

Sixth Schedule did not satisfy the tribal people. The first to take to the path of violence were the Nagas. But it was impossible for them to prevail against the Indian State. The Naga people realized after some years the futility of violence. In 1962 the State of Nagaland was established and constitutional contrivances devised to satisfy the aspirations of the tribal people were gradually integrated in the architecture of the country's power structure. There was peaceful emergence of Meghalaya (1971) and Arunachal Pradesh (1986), and the not so peaceful creation of Mizoram (1986). The three hundred and seventy-one Articles protected tribal and regional interests and identities. Even after the separation of these areas Assam is still not a homogeneous State. There is the Bodo agitation. The State of Assam includes the Bengali-speaking Barak valley, artificially attached to Assam by the British, and this has increased the proportion of Bengali-speakers in the State. The North Cachar Hills district is also not satisfied with its lot. Economic development, growth in communication, population explosion and internal migrations in search of livelihood greatly complicated the situation in Assam.

The powers of the Central Government are not derived only from constitutional provisions. The constitutional powers are, of course, quite formidable. Article 356 of the Constitution has been the most misused device. It has been invoked by the Centre for partisan purposes to topple Governments belonging to the Opposition parties. Since the Congress has been the ruling party in New Delhi most of the time, the Article has been used more often agains the non-Congress parties. But the non-Congress Centre has not been wholly innocent. The Centre can act on its own because the wording in Article 356 is: 'on a report by a Governor or otherwise'. In 1977 the Janata Party Centre asked the Congress Governments in several States to resign and hold fresh elections on the ground that the recently-held Lok Sabha elections had shown that the Congress had wholly forfeited its mandate. There was also a more solid ground, namely that their original five-year term was over (it was the Emergency Forty-second Constitution Amendment which had increased the term to six years). But the Janata Party did not use this argument for it also wanted to annex UP and Orissa which had *not* completed their five-year term. Morarji Desai's Government dismissed the Devraj Urs Ministry (of the Congress-I) in Karnataka on the eve of the 1978 poll, although Urs had shown readiness to face the Assembly and prove his majority.

In the conditions that prevail in India it would not be possible to abolish Article 356, but some curtailment of this power is absolutely necessary.

Another area of conflict is the division of resources. The States complain that the Centre takes away a bulk of the resources and starves the development activities in the States. A similar complaint is made by the local self-government institutions against the States. But the real culprits are the expansion of the bureaucracy, proliferation of ministries, and unlimited growth in administrative, non-development expenditure. The Public Sector Undertakings are also a bottomless pit into which scarce resources have been poured continuously. Private industry renders units sick thanks to mismanagement or loot and the sick units also become a charge on the State. The rich States are selfish and want the income generated by them to be allocated to them, but the backward States look to the Centre for assistance. The Centre cannot ignore their claim.

There are strong tendencies which tend to expand the powers of the Centre. There are powerful pressures which relentlessly push the Centre into accepting new responsibilities. The Akalis are the loudest in their claims for State autonomy. But to protect the interests of the large Sikh landholders in the Terai region of UP, they demand that the Centre interfere in the agrarian legislation of UP — a State subject under the existing constitutional arrangements. For long years, in the twenties and thirties, the Muslims were the most vocal champions of provincial autonomy. But now they ask for central intervention as in Ayodhya — where the Babri-Masjid-Ramjanmabhoomi issue led to a social and political crisis — and want the Centre to deploy its own para-military forces, and even the Army, wherever there are communal riots. They totally distrust the State Governments controlled by the State elites. The Harijans and the tribal people also invoke Central intervention against the locally dominant caste groups.

In most of the States local vested in erests have combined with the contractors and profiteers to destroy recklessly the forest wealth. Already India has lost most of its forests. The local populations are helpless. The environmentalists loudly call for control over the economic development in the constituent States. The proliferation of the Planning Commission bureaucracy was a development not visualized by the Constitution.

Article 301was intended to make India one vast free market area by declaring that trade, commerce and intercourse throughout the territory of India shall be free. But in fact commerce is not free. Food shortages call for imports of foodgrains and edible oil, and its allocation to the States is under Central control and direction. The public distribution system on which the urban poor — and to an extent the rural poor — depend is geared to the Centre's policy.

Another important source of centralization was the control exercised by the ruling central party leaderships. Although the Congress Central Parliamentary Board always enjoyed great authority till the last years of the Nehru dispensation, the Congress State Legislature Parties enjoyed considerable freedom to elect their leaders. In 1963 some Chief Ministers, chosen by their CLP's, were, for the first time eased out under the Kamaraj Plan. When Indira Gandhi established her supremacy in 1971 the procedure of democratic selection of leaders in the States was done away with. It was Indira Gandhi — and later Rajiv Gandhi — who picked out the Chief Ministers and removed them at will. Under the Desai regime (1977-9) more subtle means were used to unseat State Governments. Dissidents received open encouragement from Central leaders and by forcing the Chief Ministers to obtain confidence votes from time to time, several Governments were toppled. Both the Congress method and the Janata Party method had a debilitating effect on State administration.

While there is danger of an overwhelming strong Centre overshadowing the States, not enough attention has been paid to the problem of the States eclipsing the organs of local self-government. The Indian political parties pay lip service to the ideal of political devolution. But the Constitution does not define the powers, resources and mode of election of local bodies. The Rajiv Government made an attempt to secure passage to two Bills on decentralization. But they contained some clauses which aroused the suspicions of non-Congress parties about central interference in the work of local bodies at the expense of the States. The State Governments (barring a few exceptions) have on the whole treated the local bodies shabbily.

The abuse by the Centre of Article 356 in relation to the States has its counterpart in the worse abuse by the States of their powers over local bodies. The State Governments have suspended municipal administrations with impunity, postponed elections — or have not held them at all for years together — and have introduced the unhealthy practice of nomination in the local bodies. Some States allow MLAs and MPs to sit on these bodies who outshine the local representatives. The District bureaucracy also lords it over the elected local bodies. The downgrading of the local bodies, their penetration by criminal elements and the contamination of the election process at the level of local self-governing institutions is a great shortcoming of Indian democracy.

The deterioration of the State administrations, corruption and inefficiency of the State security forces, breakdown of law and order in large parts of the country and the pitiable condition of the administration of justice, as revealed by the statistics about crime committed, reported, registered, and prosecutions launched and convictions obtained, have resulted in the weakening of the Centre. The States of Punjab and Jammu and Kashmir, and some parts of the north-east have been under Central rule for years. But because of the dilapidated administrative structure and the growing alienation of the local people, the Central Government, with all its vast powers, has been unable to make any impression on the problem of terrorism and insurgency in these areas.

Electoral reform has been a subject of discussion these past several years. Of immediate concern is the restoration of the purity of the democratic process. How is the growing influence of money power *and* muscle power over the elections to be curbed? The holding of simultaneous elections to all representative bodies from the Lok Sabha down to the District Panchayats and Municipalities will drastically curtail election expenditure and also help political parties to train workers for filling these bodies competently. Nobody should be allowed to contest from more than one Lok Sabha and Assembly constituency to prevent unnecessary by-elections.

The BJP and some other parties demand the introduction of proportional representation (PR) and state-funding of elections. In a country of such immense diversities as India (there are over two thousand castes among the Hindus) PR would be disastrous. The list system would strengthen despotism within parties.

Most parties suffer from the vice of gerontocracy. The Communist Parties were the greatest offenders in this respect The BJP, too, is following in their footsteps. These parties have blocked all avenues to real promotion for younger elements. In sharp and pleasant contrast was the attitude of Gandhi (from the beginning), Nehru (in the thirties), Jayaprakash Narayan and Rammanohar Lohia.

Another topic of heated discussion has been the phenomenon of defections and politicians' fickle political loyalties. A Committee was set up which recommended that ministerial office and other posts of material benefit should be denied to all those who left the parties on whose platform they had been elected. Indira Gandhi sought to secure passage of a constitutional amendment which treated an adverse vote by legislators against their parties as defection and sought to disqualify them. This was really anti-dissent legislation and met with stiff opposition. The Amendment Bill was never passed. The Janata Party Government, under Morarji Desai, tried to revive the obnoxious legislation, but it also ran into rough weather and had to be withdrawn. In 1985 the Opposition was so demoralized and Rajiv Gandhi had won such an overpowering mandate that he was able to rush through an equally objectionable Bill in the two Houses in one sitting each. Not a single dissenting vote was recorded against this piece of anti-democratic legislation.

Ever since this constitutional amendment has been in force, it has played havoc with the party system and the democratic process. It has encouraged bulk defections. The Supreme Court refused to come to grips with the problem for several years because the judges were unwilling to tread on the toes of the Rajya Sabha and the Lok Sabha. The Constitution visualized the Supreme Court chiefly as a constitutional court. But unfortunately it neglects its main function. The Court and the Bar have changed their character into a predominantly appellate tribunal for adjudicating ordinary criminal, civil, industrial relations, company law, taxation and other cases which do not involve constitutional issues. Judges and advocates participate in the ugly game of granting frequent and unjustified adjournments.

No modern methods are used to group similar cases and achieve co-ordination among the different Benches.

In the matter of anti-defection litigation the Supreme Court was forced quickly to modify its earlier opinion that the Tenth Schedule, except the prohibition of judicial review, was valid. It proceeded to read down the provision about disqualifying those who defy the party whip. The Court said that the disqualification could be invoked only in cases which involve the Government's stability such as voting on a Confidence Motion and the Budget. It appears that in handing down this new opinion, the Court, instead of issuing guidelines to Parliament for amending the Constitution, took upon itself the job of amending the Tenth Schedule This is nothing but self-aggrandizement pure and simple.

The Indian Constitution is over-elaborate — a legacy of the Act of 1935 — and is, therefore, not a flexible instrument. It has undergone sixty-nine amendments within a space of forty-two years (the United States managed with only twenty-six amendments in over two hundred years).

The Ninth Schedule has given protection to two hundred and fifty-seven Acts on the subject of socio-economic change. Indira Gandhi abused the Ninth Schedule by including in it four anti-democratic Acts. The Morarji Desai Government had to undo the mischief. In a meaningless populist move the Desai Cabinet removed the right to property from the list of fundamental rights. In fact this changed nothing. It continues to be a constitutional right. The ten-yearly extension of the reservation of seats in the legislatures for the Scheduled Castes and Scheduled Tribes and the creation of new States necessitates several amendments.

The Objective Resolution passed by the Constituent Assembly on 13 January 1947, the Preamble of the Constitution, and the Directive Principles of State Policy embodied Jawaharlal Nehru's concept of a social welfare State. The new State sought to create a network of free elementary education and primary health centres in the rural areas. *But long before the slogan of privatization was adopted by the ruling groups in the sphere of industry and trade, the State-supported institutions had gone out of fashion in the area of education and health.*

Premature consumerism has taken its toll. The 'get-rich-quick' psychology has destroyed the work ethos. Large numbers are engaged in illicit money-making and crime. Equality has become a hated word. The rudimentary welfare State is in a shambles.

Some of our politicians, jurists, retired civil servants and intellectuals believe that systemic change in the constitutional arrangements is a panacea for the evils from which India is suffering. They want to replace the existing system by some form of presidential government. Under it all executive power would be vested in the President elected by direct election. He would not be responsible to the Legislature nor would he attend its sittings, although he could address it and send it messages. The Presidential Cabinet, too, would not be responsible to it. There would be no collective responsibility of the Cabinet. It would be there only to assist the President. The advice of its members would not be not binding on him. The adoption of this system, however, will not solve any of the problems confronting the Indian people and the State. It will obliterate the familiar landmark of parliamentary government and launch the nation on uncharted, stormy seas without a compass. The Indian advocates of systemic change, perhaps, want to clothe the head of the government with the power to dissolve the Legislature. This they hope would make it pliant and ensure stability. But such a system would quickly lead to an authoritarian form of government of which we had a taste in 1975-7 and would destroy the democratic freedom.

What is called for, therefore, is not any grandiose scheme of systemic change, but more inward change. This does not mean that there is no scope for minor practical reforms. All democratic institutions and authorities — President, Governors, Speakers, Chairmen, Prime Minister, the Chief Minister's Cabinet, the Press which sometimes abuse their freedom, legislatures which are often oversensitive, judiciary and the bar — all have to exercise self-restraint.

There must be accountability at all levels of government, and strong action against crime and corruption — the greatest barriers to quicker economic development. And above all, a work ethos must inform all departments of our national life.

Federalism, Sub-nationalism and the Indian Constitution

P. M. BAXI

The object of this paper is to explore the impact of sub-nationalistic tendencies on the federal scheme of the Indian Constitution. The focus will be on territorial changes that have been the result of the operation of such tendencies. For a proper appreciation of the position, it is desirable to say a few words about the nature of the federal structure in India, before examining the impact of sub-nationalism.

In India since 1947, developments in regard to territorial reorganization have been so extensive and so fast, that one may miss the social and political undercurrents to which many of these developments owe themselves. It is often pointed out that India is not only the world's most populous but also the most complex federal democracy of the world. The extent to which the federal structure has undergone change in quality and quantity, requires a study that might throw some light on the undercurrents mentioned above. The present essay is a modest attempt in that direction.

FEDERALISM AND FEDERAL MODELS

It is the Latin term 'feodus' which seems to have given rise to the adjective 'federal', the noun 'federation' and the concept of 'federalism'. At least eight meanings are attributed to this word in Latin (Lewis).[1] These meanings are: alliance, association, compact,

1 Lewis, *Latin Dictionary*.

contract, league, treaty, trust and union. The meaning that comes nearest to federalism as understood in modern times seems to be that of 'union'. But one should bear in mind that there could be many types of union and there may be differences in the strength of the bond that makes up the union from its federating units. Federations, confederations and allied arrangements, are not new. We have, in the ancient world, examples in and outside India of such arrangements. Certain features of federalism are illustrated in some of the Union of States established in Western and Northern Europe in the medieval period, notably the Federation of the United Provinces of the Netherlands, established in 1576. India had its federal element in the republican constitution of Vaishali, formed by the various Lichchhavi tribes. The history of Greece is also replete with the rise and fall of leagues, unions and confederations.

However, federalism, as understood in the modern era, is ordinarily evaluated on the basis of a given model. The model generally contemplated is one in which the constitution of the country establishes a union through a scheme where sovereignty is distributed between the union and the units. Switzerland is normally regarded as the oldest country in the world with the federal system.[2] The present number of cantons in that country is twenty-two but it started in 1281 with three cantons of the Alpine region, which formed an alliance for the purposes of defense.

TERRITORIAL CHANGES IN INDIA: 1947-1950

Territorial changes have been common to almost all federations. However, the rapidity with which they have been taking place in India appears to be unparalleled — if one leaves out the recent developments in Eastern Europe. It was in a letter written by the late Shri Jayaprakash Narayan to Pandit Nehru in 1961, that the former tried to analyse the concepts of nationalism and sub-nationalism. shri jayaprakash narayan[3] put forth the proposition that religion, race, language, history, tradition and geographical regions make up a nation. In reality, all these elements are subordinate loyalties in

2 G.R. Codding, Jr., *The Federal Government of Switzerland* (1955), Ch. 1.

3 Letter of Jayaprakash Narayan, dated 11 September, 1961, referred to by Shri Mastram Kapoor, 'Who is attacking national unity' (Hindi), published in *Navabharat Times*, 23 January 1992, p. 7.

comparison with nationalism. Real nationalism is a state of the mind in which a citizen gives priority to national loyalty and the national interest.

The organization and reorganization of political divisions in India has been a continuous process, before, during, and after British rule. Before 1947, these were mostly the result of conquest, annexation and similar political acts or events which brought about a change of political power. The period 1947-50 is marked by the emergence, rather than the breaking up, of political entities. The most notable change took place in the form of the integration of Princely States,[4] followed by the absorption of territories from French India and Portuguese India.

The integration took a variety of forms: (a) certain Princely States were absorbed in the neighbouring 'Province' of which the native State of Baroda is an example. The province of Bombay was already a part of the Dominion of India constituted under the Government of India Act, 1935 as adapted; (b) certain Princely States, small in size individually, but large in number, were merged together so as to bring into existence a new State. The States of Madhya Pradesh, Rajasthan, Saurashtra, and Travancore-Cochin, are examples. These newly formed 'States' then acceded to the Dominion of India, which was a federation, thus enriching the federation.

1950 and After

The integration that had been achieved in the period 1947 to 1950 was under the Government of India Act and was given constitutional status by the Constitution of India. It is useful to recall that the pre-1950 political map of India, consisted of Governors' Provinces and Chief Commissioners' Provinces (with considerable administrative differences between the two). In addition, a large portion of the territory of India, as envisaged by the Constitution, comprised the areas of former Princely States who had acceded to the pre-1950 Indian federation (Dominion of India), as mentioned above, or accepted their own merger in a neighbouring Indian province as constituted under the Government of India Act, 1935 (as it was in operation before 1950). These historical facts account

4 See V.P. Menon, *The Story of the Integration of the Indian States* (1956).

for the division or categorization of the units of the Indian federation under the Constitution (as originally established in 1950) into four parts — A,B,C and D. Of these, it can be emphatically stated that the States in Part B, by and large representing the areas of Princely States, were the most important from the point of view of federalism. The process up to 1950 was one of merger or absorption or accession into a larger union, rather than fragmentation or multiplication of parts. The process reflected a centripetal tendency, rather than a centrifugal one. The reason for this marked centripetal tendency up to 1950 can only be conjectured. The most plausible reason seems to be that the sentiment of nationalism that inspired the freedom struggle in India had created a national image in the hearts of its citizens. Literature, journalism, academic life, not to speak of the work of politicians of the time, created a psyche of unity, rather than one of division. The struggle against a foreign power cemented the ties of nationalism. There was no room for sub-nationalistic tendencies to thrive. It is also necessary to mention that integration was very largely achieved by the mighty genius of Sardar Patel, whose role in this direction has not been adequately appreciated. It was a bloodless process and even the princes acceded to it.

Linguistic Reorganization

Soon after the Constitution came into force, the movement for territorial reorganization started in India. The beginning was made by those who pleaded for the grant of statehood on the basis of the language of that area. The pioneering example in this regard is the formation of the State of Andhra Pradesh in 1953. Linguistic sub-nationalism thus became prominent soon after the Constitution came into force. In fact, it became the major factor for frequent redistribution of territory in the years 1950-60.

Sub-nationalism and Language

The movement for the establishment of a separate State on the basis of the language spoken by the majority of the population of the areas concerned, represents sub-nationalism through language. Of course, such reorganizations do not take place in a logical manner, nor are they inevitable; they need not also be simultaneous. But the point is that language has been a major factor in sub-nationalism in recent years.[5]

The year 1953 saw the birth of the State of Andhra. In 1956 the Indian States were 'reorganized'. This was achieved through the passing of the Constitution (Seventh Amendment) Act, 1956 and the States Reorganization Act, 1956. By the Andhra State Act, 1953, the State was later renamed as Andhra Pradesh in 1956. This was followed in 1960 by the reorganization of the State of Bombay, through the Bombay Reorganization Act, 1950. The partition of the State of Punjab into two States — Punjab and Haryana — was achieved by the Punjab Reorganization Act, 1966, while the formation of the State of Himachal Pradesh was achieved by the State of Himachal Pradesh Act, 1970.

Constitutional Provisions

It is not that territorial reorganization was outside the contemplation of the makers of the Constitution. Article 3 of the Constitution (subject to certain safeguards), specifically authorizes Parliament 'by law', *inter alia*, to form a new State by separation of territory from any State or by uniting two or more States (or parts of States), or by uniting any Union Territory to a part of any State. It also authorizes Parliament, by law, to alter the boundaries of any State or to alter the name of any State. Besides this, Article 4(1) of the Constitution authorizes Parliament, by such law, to amend the relevant Schedule to the Constitution and to enact supplemental, incidental and consequential provisions 'as Parliament may deem necessary'. Article 4(2) makes it clear, that such a law shall not be deemed to be an amendment of the Constitution for the purposes of Article 368, which is the provision of the Constitution dealing elaborately with amendment of the Constitution and prescribing certain special formalities for the purpose. These provisions of the Constitution have been judicially described as dealing with internal adjustment, *inter se*, of the territories of the Constituent States of India (A.I.R. 1960).[6] This measure of elasticity in the territorial map of India was considered necessary, obviously because the Constitu-

5 Sarbadhikari, Pradip (1991), 'India's Subnationalism' in *Canadian Review of Studies in Nationalism* vol. XVIII, nos. 1-2; Brass, Paul (1968), *Language, Religion and Politics in India* (Cambridge University Press)

6 *Re Berubari Union*, A.I.R. 1960 S.C. 845, 856.

tion-makers were aware that the necessity might arise for such redistribution of territories and that, if such a necessity arose, the cumbersome machinery of constitutional amendment provided in Article 368 would not be very convenient. Although they did not apprehend frequently recurring moves for territorial redistribution, they were aware that demands for reorganization of States on the basis of language might come from various quarters.[7]

Certain historical aspects relevant to the linguistic issue may be mentioned at this stage. B.N. Rau, who was the Constitutional Advisor to the Government of India when the Constitution was being framed, has expressed the view: 'One of the most difficult problems in the framing of India's new Constitution will be to satisfy the demand for linguistic provinces and other demands of a like nature'.[8]

The Constituent Assembly did not expressly provide for the linguistic reorganization of States, but it did make a provision in Article 3 of the Constitution, providing for reorganization in general. Incidentally, the precedent for this article was Section 290 of the Government of India Act, 1935 which, in its turn, closely followed Chapter 6 of the Australian Constitution; and the parent of all such provisions is Article IV, Section 3 of the Constitution of the United States. In the actual debates in the Constituent Assembly, the issue of linguistic provinces did come up from time to time but it appears that Pandit Nehru, Sardar Patel, Rajendra Prasad and Maulana Azad did not agree to any redistribution on a linguistic basis.[9] It is also relevant that a Linguistic Provinces Commission was constituted on 17 June 1948 whose Chairman was Mr Justice S. K. Dar, retired Judge of the Allahabad High Court. That Commission concluded: 'The formation of provinces on exclusively or even mainly linguistic consideration is not in the larger interest of the Indian Nation and should not be taken in hand'.[10] Eight days after the publication of this Report, the Annual Congress Session at Jaipur approved a resolution, forming a Committee consisting of

7 Austin Granville, *The Indian Constitution: Cornerstone of a Nation* (OUP: First Indian edition 1972, reprint 1985) pp. 236 – 41.

8 Rau, *Constitution Precedents*, First Series, p 17.

9 See *Constituent Assembly Debates*, vol. 1, pp. 5, 57; vol. VIII, pp. 4, 320, 321.

10 Linguistic Provinces Commission, Report, para 152(I).

Pandit Nehru, Sardar Patel and Pattabhi Sitaramayya to take a second look at the question. The Committee, in its report dated 1 April 1949, recorded the view: 'We feel that the present is not an opportune moment for the formation of new provinces'. Yet the members also believed that if public sentiment was insistent and overwhelming, they, as democrats, had to submit to it, but subject to certain limitations in regard to the good of India as a whole. Meanwhile the supporters of linguistic provinces continued to press their claims.

Nothing binds two or more individuals together, as a common language. The Constitution of India itself, in Article 29(1), provides that any *section of the citizens* residing in the territory of India or any part thereof, having a distinct language, script or culture of its own, shall have the right to conserve the same. It is to be noted that this provision is not confined to linguistic minorities as such.[11] Further, this right has been held to cover the right to embark upon political education for the protection of the language.[12] As regards minorities, Article 30(1) of the Constitution provides that all minorities, whether based on religion or language, have the right to establish and administer educational institutions of their choice. But it is doubtful if it was anticipated that there would be a widespread and persistent move for the reorganization of States soon after the formulation of the Constitution.

The Cultural Aspect: The North-Eastern States

Another major force that seems to account for territorial reorganization in India is what may be called the cultural aspect. This is illustrated by the formation of the States of Nagaland, Meghalaya, Mizoram and Arunachal Pradesh as a result of the North-Eastern Areas (Reorganization) Act, 1971. By this Act, these new States were carved out of the pre-existing State of Assam with certain other territorial adjustments. The State of Nagaland had been earlier established in 1962. As regards the States of Manipur and Tripura, the areas concerned were already being administered under the respective names as if they were Chief Commissioners' Provinces. What the North-Eastern Areas (Reorganization) Act, 1971

11 *D.A.V. College* v. *State of Punjab (I)*, A.I.R. 1971 S.C. 1731, paras 5, 6, 12, 18.

12 *Jagdev Singh* v. *Pratap Singh*, A.I.R. 1955 S.C. 183, 188; and *Rama Rao* v. *Telugu Desam*, A.I.R. 1983 A.P. 96.

achieved was merely their elevation to the status of 'States', instead of their earlier status of Union Territories.

The reorganization of the States in the north-east represents the recognition of the separate cultural identity of the population of the respective area; in some cases, there may also be a linguistic identity, but that does not seem to have been a major force in the move for the grant of a separate State.

The reason why the feeling of sub-nationalism in the States in north-eastern India acquired strength during the last decade or so is historical, cultural and linguistic. The following circumstances appear to be relevant in this context: (i) The areas of these States are mostly mountainous and are thus geographically different from the areas covered by the Great Indo-Gangetic Plain: the environment of the hills may possibly lead to a different attitude amongst the inhabitants; (ii) the majority of the population in these areas seems to be living under customs basically different from those prevalent in the rest of India: for example, in the State of Meghalaya, the Khasis, Garos and certain other tribes follow the matrilineal system of succession and property, which is different from the system followed in the rest of India (excepting certain communities from Kerala); (iii) the linguistic structure is somewhat different from the rest of India, or at least from northern and eastern India, where a large part of the vocabulary is derived from Sanskrit; (iv) the areas in question were not brought under the direct sway of the Mughal emperors or other empires that have ruled a large part of the territories within India at various periods of history.

Most of the propositions enunciated above apply to the north-eastern States, with the possible exception of Assam and Manipur. Even in these two States, the remoteness of the areas from the mainland might have been responsible for the phenomenon of sub-nationalism overtaking nationalism. One has also to take note of the fact that some of the areas in question are situated at the very frontiers of India and there is a possibility of cross-fertilization of Indian culture with 'non-Indian culture'.

SOME TENTATIVE HYPOTHESES

What could be the possible reasons for the assertion of sub-nationalistic tendencies in India during the last few years? To answer this question, one has again to turn to the proposition put forth by Shri Jayaprakash Narayan. When one or another component of nationalism becomes so predominant as to overpower nationalism itself, then sub-nationalism triumphs and nationalism recedes into the background. It is not that India cannot be viewed in unity. Historically, if one looks at the immediate pre-Constitution picture, India was already a federation under the Government of India Act, 1935. It is true that up to 1947, the Princely States did not actually accede to the federation, but the Act of 1935 definitely contemplated such accession. In any case, the Provinces were already functioning as units of a federation. In fact before the Act of 1935 and under the Government of India Acts of 1915 and 1919, British India was a unitary State. The 'devolution' of power that took place under those Acts did not attain the status of a formal constitutional division of sovereignty because the devolution could be made or marred by the Centre. It is worth pointing out that, in fact, the constitutional and legal phraseology employed under those Acts was 'Central Government' and 'Local Government'. Thus, the political setup for the maintenance of unity of the country was operative under the Government of India Acts, 1915 and 1919.

The present Constitution is not a 'State-oriented' one. In fact, one of the complaints made by foreign writers in regard to the Indian Constitution is that there is a tilt towards centralization in the constitutional scheme. Some writers have even hesitated to describe the Indian Constitution as a federal one. Professor K.C. Wheare had to coin the word 'quasi-federal' to describe it.

It is true that before the British came on the scene and established their sway throughout India, the country was governed by a large number of kings, princes and vassals, functioning under various names and possessing political independence in large or small measure. Some of them even had their own modes of international relations. But, during the ancient period, a national consciousness must have been prevalent to a large degree. Moreover, at various periods of history, great empires encompassing a very large portion of what is known today as India, existed and thrived for centuries.

Thus, the conditions requisite for fostering the spirit of nationalism in India have existed from time to time. If sub-nationalism has gathered strength notwithstanding their existence, the reason lies in the fact that political currents and campaigns have given a secondary place to nationalism and brought into greater focus the sub-nationalistic tendencies. These tendencies were lying dormant in the psyche of the inhabitants concerned, and it did not require much effort to provoke them, to magnify the sub-national sentiment, and ultimately to make it conquer the unifying spirit of nationalism. The human mind is so constituted that one can compare it to a vast and deep mine of so many minerals. Some of the minerals may be beneficial and full of lustre, while a few may be dark and malevolent. If the mind remains passive and allows itself to be overpowered by the forces that deliberately strive to bring out the darker elements, they will conquer the psyche. This is true of individuals. But it could be much more true of groups of individuals. In a multitude, reason takes a back seat when passions are inflamed and emotions come to the forefront. This process is encouraged by a strong-minded leader acting, with all his rhetoric, on the minds of persons with weak intellects with minimum powers of resistance and with their sense of judgement rendered ineffective. The minds of the multitude come down to the lowest common denominator of evil. They will not ascend to the highest common multiple representing the best and the most excellent qualities. In other words, unless there are countervailing beneficial forces which try to maintain the national sentiment, fissiparous tendencies based on language or culture might take over. It is in this sense that sub-national sentiments based on religion, language, culture, tradition and the like, tend to capture the minds of people, making nationalism a victim in the process.

State Power and Indigenous Capitalist Development in India

BIPLAB DASGUPTA

Socialism was, until very recently, a popular word in India. The constitution of the country, as also the plan documents, declared socialism as their goal, while practically every political party, including the Congress and the BJP, affirmed their allegiance to some form of socialism. The word socialism was never defined, excepting in the vaguest possible manner: some seeing it as more than an element of redistributive strategy in the plan framework; some identifying it with the public sector; while some others taking it as no more than the 'trusteeship' concept of Mahatma Gandhi, according to which the rich, while doing everything possible to become richer, were expected to look after the interests of the poor and to make some sacrifice in order to avoid class conflict and tension.

In recent months, against the background of the economic crisis which led to fresh loan agreements with the International Monetary Fund, subject to conditionalities, the issue has been posed whether 'Nehruvian Socialism' had failed in India, and, hence, whether it should be abandoned in favour of a development strategy which is biased towards the private sector, market-based allocation of resources, goods and services in response to price signals, liberalization of imports and promotion of exports in place of import substitution, and abandonment of redistribution based on extensive subsidies, direct or hidden. The new industrial policy, declared in the second half of 1991, after showering praise on Nehru for his 1956 policy in the first paragraph, indicates a radical shift from his policy in the rest of the document. This shift is justified on the ground that times have changed. The collapse of the Soviet Union has strengthened the argument that a 'command economy', sheltered by protectionist policy and stringent import control will not work, while the most urgent priority for India is to dismantle walls around the national economy and globalize.

I would take the view that the path pursued by Nehru was not a 'socialist' one: on the contrary, his objective was to build capitalism, and it failed because the Indian State was unable to fulfil the necessary conditions for such indigenous capitalist development, e. g. implementation of land reform. Further, the very fact that Nehru, and later Indira Gandhi, were unable and unwilling to fulfil such conditions, is largely because of the character of Indian State power where indigenous capitalists, strong at the national level, are dependent on the kulaks and other remnants of the feudal system at the grassroots level for holding political power. This uneasy alliance, with all its inherent contradictions, between the rural rich and the indigenous capitalists, rules out comprehensive and radical land reforms without which neither agricultural productivity can be raised to produce enough surplus for industrialization nor can the market for industrial goods be widened within the country, to take full advantage of the economies of scale and sustain such indigenous capitalist development.

THE NEHRUVIAN MODEL

The Nehruvian model was identified with 'socialism' because (a) it relied on the public sector; (b) for building the public sector it took the help of the socialist countries; (c) the development of the economy was supposed to be undertaken within the framework of a planned economy; and (d) from theAvadi Congress of 1955, the ruling party declared its commitment to a 'socialistic pattern of society': redistribution of wealth and eradication of poverty being two of the major goals of the planned economy. This 'socialist' identification was further strengthened by its foreign policy, based on the principle of non-alignment, its anti-imperialist stand on a variety of issues along with some other countries of the Third

World, in particular Egypt and Yugoslavia, and the frequent visits by socialist luminaries, such as Khrushchev and Chou en Lai, particularly on the eve of elections. Let us examine these four elements in turn.

Public Sector

In India the public sector was not meant to compete with the private sector; its role was supplemental. Though some private companies were taken over, e. g. some jute mills, the coal mines, airlines, etc., it was due, almost invariably, because the private sector was running at a loss and was incapable of overcoming those problems. The main thrust of the public sector was to set up industries in those areas where, for a variety of reasons, the private capitalists were unwilling to invest — the 'commanding heights' of the economy which influenced the course of development, and these also with long gestation periods, enormous capita, and high risks. Although indigenous capitalists wanted such industries to be set up within the country, they did not wish these to be under the control of foreign enterprises; but they themselves did not venture to set up these. The public sector was not viewed by them as a threat as long as the State eliminated foreign dominance, or control by any one Indian capitalist group at the cost of others. The public sector was prepared to undertake the massive investment that was called for, to wait for a very long time before the investment yielded profit, and was prepared to take the risk of loss, but at the same time helped to create the infrastructure for the development of indigenous capitalism.

Socialist Support

The main support for building the public sector — by way of credit, technology and construction of 'turnkey' projects — came from the socialist countries, though it was not originally intended that way. The Indian Government sought the support of Western countries and the World Bank, in the first instance, but was refused and, therefore, had no option but to turn to the socialist countries. Eugene Black, the Chairman of the World Bank at that time, was particularly severe on India for opting for the public sector; his view was that India's comparative advantage was in agriculture while the industrial goods should be mainly imported and not produced within the country, and whatever industry was needed should be set up with the help of multinational companies that had the necessary expertise and resources.

By opting for the public sector, Eugene Black thought, India was squandering scarce resources and consequently did not deserve any help. It was because of this reluctance on the part of the West to help that India turned to the Socialist countries. The latter provided credit at a cheap and highly subsidized rate of interest, technology without preconditions, and left the industry concerned in Indian hands without insisting on ownership. The Socialist help was important also for two other reasons. First, in the field of oil exploration and refining, the foreign oil companies operating in India from the 1880s never showed much interest in exploration, and always gave the impression that India was poor in oil resources; but the Soviet experts, almost immediately after coming to India in the early fifties, found oil in Nahorkatiya and Ankleshwar, identified a huge prospecting oil-bearing area spread over many States (of which only a fraction has so far been explored) and then went on to help with the setting up of the ONGC. They also helped to make India largely self-reliant in refinery products, by setting up public sector oil refineries and providing crude oil to supplement domestic production. Secondly, more important than the actual help given was the fact that the socialist option existed, which tremendously strengthened the bargaining position of India vis-à-vis the western countries; they could be depended upon for help when it was not available elsewhere. Even at the height of Indo-Soviet co-operation, the Soviet Union accounted for less than 10 per cent of the total foreign loan, less than 9 per cent of total imports, and less than 20 per cent of total exports. Imports comprised critical inputs and exports consisted mainly of non-traditional goods for which there was no market in the developed West. It was with Socialist help that India became largely self-reliant in defence, refineries, and many other important lines of production. We should also note that such Socialist support was not conditional on India's following a socialist path internally; in fact such help was made available to many other Third World countries which had no socialist pretensions. To the extent such help made the Third World countries less dependent on Western help and strengthened their resolve to follow a policy of non-alignment, it helped the Socialist countries by making the world less hostile towards them at a time when they were being increasingly encircled by military pacts sponsored by the United States.

Plan

Though the concept of planning was first introduced to the world by the First Soviet Plan in 1928, in India, even during the pre-Independence days, big businessmen deferred formulating plans for self-reliant development. These plans took shape only after Independence, e.g., the Bombay Plan. The First Five Year Plan was launched in 1951, a few years before socialism became an integral part of public rhetoric. What has given a 'socialist' identification to Indian plans, is the emphasis on heavy industries during the Second Plan under the leadership of Professor P. C. Mahalanobis, which led to the development of steel, machine-building refineries, and other industries. The reservation of these major sectors of the economy under the public sector, in the Industrial Policy Resolution of 1956, also helped to strengthen the 'socialist' identification, though, as stated above, such public ownership was necessary for indigenous capitalist development. The 1951 Industrial Development and Regulation Act, which regulates location and licensing of private industries, also predates socialist rhetoric, though later on, this legislation was viewed as a key one for creating a 'command economy'. In fact, planning was never taken seriously in India. The data system for such planning was always lacking, as was also the infrastructure for implementing it. There were occasional 'plan holidays', and often yearly plans were finalized at the fag end of the year and, therefore, these made no sense. It was J.K. Galbraith, then United States Ambassador to India, who once said that if India were socialist the United States was socialist twice over, should socialism be measured in terms of control exercised over the economy by the Government or in terms of correspondence between Government policy decisions and their implementation.

Rhetoric

Though socialism became an integral part of public rhetoric from the mid-fifties, it was never defined, excepting in terms of public ownership of some of the key sectors of the economy and the formulation of plans. During elections, the ruling party made full use of the complimentary remarks made by the Soviet dignitaries during their Indian visits. That was a time when socialism emerged as a strong ideological force the world over, with one-third of humanity living under the socialist rhetoric. It was a way of neutralizing the growth of political forces genuinely committed to socialism within the country. Such use of socialist rhetoric was made by leaders of many other Third World countries, in order to maintain their political control and to build indigenous capitalism with the public sector as a major instrument, while undertaking repressive measures against Communist Parties within the country at the same time. *'Garibi hatao'* (banish poverty), 'growth with social justice', 'redistribution with growth', 'poverty-focused planning', were some of the jargons used all over the Third World, often under the leadership of the World Bank, partly to conceal the real character of State power to confuse the masses, and partly to provide some relief to the poor and oppressed in order to keep inequality, unemployment and poverty within tolerable bounds.

But why did this particular model of indigenous capitalist development fail? The most important explanation, as we have already stated, is the failure to carry out land reforms, a pre- condition for full capitalist transformation. In the West such transformation took place over many years, often leading to violent conflicts with feudalism, e. g. during the French Revolution or at the time of the Cromwellian take-over in England. But no developed capitalist country could avoid this step and still develop. In the East, in Japan, the process of capitalist transformation began with the Meiji Restoration of 1868 and ended with the drastic land reform under the military regime of General MacArthur after the defeat of Japan in the Second World War. In Korea and Taiwan such land reforms were carried out by the Japanese imperialist rulers themselves for their own reasons — to curb feudal military power and to secure a steady supply of rice, from a modernized agricultural system in these dependent countries to food-deficit Japan.

In India's case, land reform was on the agenda of the Indian National Congress during the pre-Independence days when the zamindars were opposed to the national struggle for independence and sided with British rulers; but, after Independence, these feudal elements were incorporated into the ruling party, as they wielded a great deal of power and influence in the countryside. However, partial land reforms were carried out in the fifties, which led to the abolition of large-scale landlordism (though it still survives in Bihar, MP, Rajasthan and some other States on a reduced scale) and the transfer of power in the countryside to small landlords and rich peasants. The land-reform legislation contained many exemptions and loopholes which permitted the landlords to undertake large-scale evasion and retain a good part of the land. The rounds of legislation in the seventies removed many of those loopholes and were quite radical and progressive, but these were mere showpieces, to be presented to world gatherings on agrarian reform in Rome and elsewhere, and were not meant to be implemented (excepting in West Bengal and Kerala, under the auspices of the Left-wing governments).

The power and influence of this particular class of rural rich was augmented in various ways. The adoption of adult franchise meant that the rural rich, with their control over rural masses by way of the patronage system (known as '*jajmani* dependent system' in India), could not bargain with their political support for various concessions in exchange for political support during elections — practically all the political parties in India, excepting the Communist Parties, depended on them for votes in the rural areas. Rural co-operatives also came under their control, as also rural panchayats, the village-level elected bodies for decentralized development, while their influence over the village administration at the community block level enabled them to divert state resources — in the form of loans, grants and subsidies — meant for the poor and the weak, under various poverty eradication measures, by manipulating the records.

The introduction of 'green revolution technology', based on high yielding seed varieties, modern chemical inputs like fertilizers and pesticides, and machinery such as tractors and irrigation pumps, further reinforced their control over the village society. They were the first to adopt this technology, to take advantage of bank loans to the agricultural sector after nationalization in 1969, and to increase substantially their productivity and surplus for marketing. While this development indicates a certain amount of capitalist development in agriculture, this took place within the framework of traditional relations of production, as land reforms had not been implemented. In fact, the success of the new agrarian technology created a new type of dependency of the poor on the rich — for high yielding seeds, for access to new technology and information, and for the hiring of tractors and irrigation pumps and, sometimes, for the purchase of controlled water without which this technology would not work. In many cases, the small farmers, not having enough capital to make the most of this new technology, leased out land to rich farmers, thus creating a new type of landlord-tenant relationship. The success of the 'green revolution technology' in Punjab, Haryana, western UP, the Krishna and Godavari delta of Andhra Pradesh, and some other places has led many to argue that land reforms are no longer necessary, without realizing that a proper capitalist transformation is not feasible without land reform.

At the same time, outside those pockets of capitalist development, there are vast stretches of countryside where the green revolution technology has not worked and cannot work because of lack of water. In those areas, which constitute the bulk of the Indian countryside, semi-feudalism of the worst kind still prevails. It is not a matter of coincidence that the four States — Bihar, UP, Rajasthan and Madhya Pradesh — with the highest incidence of atrocities against scheduled castes and women, are also those where the record of implementation of land reforms is the worst. Nor is it coincidental that the worst form of communalism and divisive tendencies exist in those States. Capitalist development presupposes the existence of secular values, literacy, and a modern attitude towards women, which are conspicuous by their absence in a large part of the countryside, largely because of the failure of implementation of land reforms.

There is still another major reason for the failure of the Nehruvian model of indigenous capitalist development, which has a lot to do with the direction of industrialization. In the industrialized countries of today capitalism thrives by producing articles of mass consumption, while in India a very significant part of the capital, foreign exchange, skilled manpower, and modern technology is deployed to satisfy the demands of a small proportion of the Indian population who have the means to purchase. This elitist section, though small in proportion to the total population, say two per cent, accounts for a very large population of one crore seventy lakhs, going by the population size of the vast majority of the countries in the world. While the remaining 98 per cent are unable to satisfy their needs because of lack of purchasing power, and cannot stimulate the Indian capitalists to produce for them as they cannot emit price signals by backing up their need with money, the elitist 2 per cent have the means to influence the market as they are capable of translating their needs into what we economists describe as 'effective demand'. The present economic crisis in India has been mainly due to the policy decision taken under the Rajiv Gandhi regime to satisfy the demand of the elite for 'white goods', by creating domestic industries of fridges, washing machines, VCRs and TVs, on the basis of commercial loans from abroad guaranteed by the Reserve Bank of India. Such an elitist orientation in industrial production, cannot be avoided. But it limits the scope for industrial development beyond a point because of lack of market without a deliberate policy of doubling purchasing power mainly by way of land reforms, by which a vast market of 850 million people could be created for articles of mass consumption.

A third explanation for the failure is the way the policy of import substitution has been promoted in India. There is nothing wrong with the policy as such — that is to provide temporary protection from foreign competition to those industries which can ultimately stand on their feet — because that is the only way to create an industrial base in a backward country when technologically and otherwise the foreign competitors are much stronger. Such a policy provides to some of the indigenous industries a breathing space without violent competition from much stronger foreign rivals, and this has been successfully followed by the newly industrialized countries in the Far East in recent years. Indeed, from the point of view of a backward country which was under colonial domination for a long period, it is not easy to follow an alternative policy of export-led growth, without first developing domestic industries catering to the domestic market (small island countries like Singapore or Hong Kong for whom there is no alternative to export-led growth are an exception to the rule). It is only after a company has been firmly established in the domestic market, patronized by the State, that a venture to take on competitors on foreign soil can be tackled. What went wrong in India was the way such a policy of import control was administered to boost the profits of major domestic companies — like the cars produced by the Birlas — while doing practically nothing to make them efficient and, eventually, competitive and strong enough to withstand foreign competition in the domestic market. Furthermore, controls, whether on imports, location decisions, or investment, and the accompanying licensing system, led to widespread corruption, nepotism and bureaucratization of the system, which made the Indian industries even less competitive over time.

All these have given rise to a situation where India, a country which is richly endowed with natural resources — whose skilled manpower is more than adequate for the purpose of industrialization and whose population size permits enormous economies of scale — and large-scale introduction of modern technology, now ranks as one of the poorest twenty countries in the world. But, as we have tried to show, the root of the problem lies in the failure of the indigenous capitalists to sever their links with the rural rich and their inability to carry out land reforms because of their political dependence on the latter.

Since July 1991, under the new regimes, a determined effort has been made to 'globalize' the Indian economy, under the conditionalities imposed by the IMF and the World Bank, by dismantling import control and licensing; by making the currency; convertible and carrying out devaluation; by inviting the multinationals; and by undertaking privatization of companies that are currently under public ownership. The current official orthodoxy is to rely on the market for allocation of goods, services, capital, and labour power and to be guided by the price signal in formulating economic policy. 'Liberalization of the economy' is the catchword. A very romantic view of the market is being projected which predates the 1929 world economic crisis, and which has little in common with the economy of the rich countries of today, where the State continues to play a leading role in formulating economic decisions.

The Indian economy is being opened up at a time when the world economy is contracting and markets are shrinking, and when the United States itself, the main proponent of such a policy, is adhering to quantitative restrictions on imports of textiles under the multifibre agreement. Devaluation has been carried out without taking into account the low elasticity of Indian imports and exports. Nor is it easy to keep import of essential goods in check without provoking inflationary tendencies — more so as the domestic production of oil is declining, and there is every possibility of world oil prices going up. On top of all that are the Dunkel proposals, or GATT rules, which would destroy the option of independent scientific and technological development; had these been implemented thirty years ago, Japan's development would have been seriously curbed.

It appears that the present regime is throwing away the baby with the bath water, dismantling a system of control and closing the option of self-reliant development, without which indigenous capitalism can never grow. Nor can the process of modernization in agriculture, initiated by the technology of green revolution since the mid-sixties, be sustained without subsidies for inputs, without support for agricultural prices, without an extensive network, and without rural banks for providing loans on a priority basis to agriculture, even at the cost of low profit to the banking sector. There can be no indigenous capitalist development in India under IMF conditionalities.

Lastly, it might be asked why the Indian capitalists are so euphoric about the new economic policy if it is so harmful to their interests? This is probably partly because of the belief that in any case not much of foreign capital would come India's way : Germany would be preoccupied with its eastern part and the neighbouring East European States, Japan would continue to operate in South East Asia and China, while most of US capital would go to the erstwhile Soviet Union. There is also the view that the dismantling of controls and associated bribery would ensure a saving on their account which would more than compensate for any difficulty they would face with foreign competition. Some indigenous capitalists take the view that after all the world today is dominated by a single superpower and it is in their interest to collaborate and get their crumbs while some are indeed shaken by the liquidation of the socialist option in the world market. But at the same time the present regime, despite its open door policy, has managed to project the impression that it is on their side and would not sacrifice their interest. The picture will not be clear for some time, but it is possible that in the course of the next few years, as the full implications of these policies of capitulation unfold, some opposition to such policies would develop from the ranks of Indian businessmen. But whatever be their views on an open door policy, there is still no indication of a sincere commitment to land reforms, and a willingness to fight the rural rich to its logical end, without which there can be no indigenous capitalist development.

POSTSCRIPT

By now the Indian economy has completed a few years under the New Economic Policy, and there is sufficient data to indicate which way the economy is moving. The GDP growth has been the slowest in recent years — 1. 2 per cent in 1991-2 and no more than 4 per cent in 1992-3 — compared to an average of 5. 5 per cent in the eighties. Even that growth has not been due to progress in agriculture or industry, but largely because of a dubious growth in services. The figure of food production of 176 million tons in 1992-3 is identical with that for 1990-1, which implies a decline in per capita food output. Industrial production actually declined in 1991-2, and the growth rate during 1992-3 was not expected to be more than 4 per cent, assuming that the riots have taken a heavy toll, compared to an average of 8 per cent during the eighties. Despite the devaluation of 1991 and partial convertibility of the Indian currency (with further implicit devaluation) in 1992 the trade deficit has actually widened to $ 7b. in 1992-3.

India would need a very high rate of growth of exports in the coming years to eliminate this gap; the composition of export trade, with gems and jewellery as the leading item, and highly competitive garments as the second, does not give much hope of a dramatic increase in export earnings.

While the foreign exchange reserve in the country has indeed improved, from around Rs 2,500 crore to around Rs 16,000 crore, such an improvement is spurious and has not come about by way of increased export earnings. The three main contributors to the foreign exchange reserves are: (a) foreign loans from the IMF and other agencies, which would have to be repaid; (b) immunity schemes which helped to convert a part of the Indian black money into legal import of foreign exchange; and (c) the short-term deposits of NRIs, which can be withdrawn at any time if the financial conditions in the country deteriorate or if the foreign countries offer better premiums. The main success of the Indian Government over the past few years has been in reducing the fiscal deficit, from a figure of 8.4 per cent of GDP to around 5 per cent now, which has also helped to bring down inflation. However, the method adopted to bring down the deficit has also, indirectly, contributed to the crisis in the external account, i.e., by severely deflating government expenditure on agriculture, rural development, etc., and private expenditure, by way of higher interest rates. The alternatives of reducing the deficit by increasing income — by seizing Rs 60,000 crore of black money (if not the stock, at least the annual flow), by imposing agricultural and wealth taxation on the rural rich, and by making the industrialists pay more taxes — have not been explored. As a consequence, while in the USA and the UK taxes on income, assets, and profits account for about half of the tax income of their governments, in India these sources account for less than 20 per cent of tax revenue, while the remaining tax revenue comes by way of regressive, inequitable and inflation-generating indirect taxes. Here again, the issue of 'political will' is more important than the economic feasibility of raising the required revenue, since such policies would offend those who are the main backers of the ruling party.

Confining ourselves to the issue of macroeconomic management, we find that the present set of policies would not succeed in handling both types of deficit, internal and external, at the same time. If the emphasis is on reducing internal deficit by adopting a deflationary policy package, this curbs export earnings as the capacity of the economy to produce exportable goods drops, and also generates demand for further imports, leading to a still bigger deficit in the external account. This dilemma has been faced by all the Third World countries which have relied too heavily on the IMF for rectifying their balance of payment problems and have faithfully followed their conditionalities over the past twenty years since the oil crisis of 1973. One cannot cite the evidence of a single country which, having subjected itself to such conditionalities over a fairly long period, has succeeded in ironing out its external and internal economic difficulties and has managed to get out of the debt trap. The experience of countries with problems similar to India's, e.g., Brazil, Argentina, Mexico, Nigeria, has been more or less uniform, moving from small loans and a small number of conditionalities to much bigger loans and more and more stringent conditionalities, losing them economic independence, swapping the shares of their domestic companies for such debts under the auspices of the World Bank, and thus virtually handing over their economies to US multinationals. Brazil, a country no less fertile than India, but three times bigger in area and with one-eighth of India's population, and reckoned as a middle-income country, has also failed to get out of the debt trap; in fact poverty and unemployment are mounting, though some success has been achieved in curbing inflation in recent years. If that is the fate of Brazil, how can India hope to do any better?

Already the debt-service ratio has reached a thirty per cent figure, much above the 'danger line', and the total external debt is three times bigger than the GDP. Given this, and the continuing deficit in external account, there is no doubt that at the turn of the century, when all the accumulated loans would have to be paid off, India would be forced to take another loan to repay those, whose figure would be so high as to make India the biggest loanee in the world, topping Brazil and Mexico who now hold the first two places. We are already deep inside this debt trap, and the policies of 'globalization' 'liberalization', etc. , being followed by the Government now, throwing all caution to the winds, would make us move deeper.

The path of export-led growth does not exist for India. India has no colonies or established trade links, nor has it an economy which is diversified and productive enough to surge forward with dramatic speed in the world market by pushing out the competitors who are already in occupation. Nor is the Indian corporate sector a match for the giant multinationals of the West, some of whom own assets, the values of which are close to two-thirds of the Indian GDP. The 'open door' policy would certainly help the latter in a shrinking world market, but it is not at all clear how the Indian companies would benefit from such open competition. Nor are the expectations that the multinationals would bring in a great deal of capital, arrange for the export of Indian goods, and provide technology likely to be fulfilled, as a quick reading of the experiences of the Third World countries the world over with these multinationals would show.

It might be asked: What factors led to the current economic impasse? The answer is not simple. It is clear that a crisis of this magnitude cannot be the outcome of policies implemented over the past few years; one has to take into account the economic policies pursued over the past four and a half decades, ever since the country became independent. The potential of a country with a huge home market, a large reservoir of skilled manpower and abundance of natural resources was never properly tapped. As has been stated earlier the market the Government aimed at was that of the urban and rural elite with purchasing power: production of mass consumption goods was never given the priority it deserved. After the accession to power by Rajiv Gandhi, even greater emphasis was given to production for the elite with foreign collaboration and with extensive foreign commercial loans. The foreign commercial borrowing increased from a figure of less than Rs 1,000 crore to Rs 26,000 crore during this period, the total foreign borrowing quadrupled. While in 1984 the quantum of foreign loan, at about Rs 27,000 crore, was manageable, within five years it crossed the Rs 1,00,000 crore mark and the country faced an unprecedented economic crisis when even commercial lenders refused to give further loan to India as the credit rating of India had reached an all time low — India became the third biggest borrower after Brazil and Mexico, and the twenty-first poorest country in the world. While this loan became necessary to meet the needs of the elite comprising two per cent of the population, its burden is breaking the back of the other 98 per cent. It is paradoxical that whereas the capitalists in the developed countries of today grew mainly by producing items of mass consumption and capturing the mass market, their counterparts in India have failed to learn from that experience and are mainly concentrating on the richer two per cent having a massive share of the aggregate purchasing power.

In the Second Five Year Plan, an attempt was made to follow the path of independent, self-reliant capitalist development by way of import substitution, establishment of basic publicly owned industries and technical and financial support from the socialist countries. Where the plan model, developed by Profesor P.C. Mahalanobis, went wrong was in its lack of emphasis on land reform in particular and agriculture in general. How could there be full-scale industrialization without tapping the vast rural market — by diffusing rural purchasing power — unleashing independent peasant initiative, and bringing into being a network of agriculture-based industries? This deficiency was exposed soon: the food crisis of the late fifties and the early sixties led to a serious foreign exchange crisis, which forced the economy to depend on the United States for both PL-480 food assistance and also for foreign aid. This more or less finished off planning as a serious exercise in India, with a long-term objective to industrialize the economy. Such industrialization was achievable only if there was a political will to carry out drastic land reforms.

This was essentially a political question — whether the ruling party at the Centre was bold and confident enough to risk antagonizing its power base in the rural areas.

As a consequence, what emerged in place of proper planning for building independent State-directed capitalism, was an elaborate system of controls and permits fed by nepotism and corruption. The main beneficiaries were bureaucrats and some of the leading capitalists, e. g. the Birlas, having close links with the ruling party, who, sheltered by control, made little effort to improve efficiency and productivity of their firms, but, nevertheless, continued to am huge profits.

There can be no question of justifying the old policies which brought the country into this mess. Something new was needed. But anything 'new' is not necessarily better than something 'old'. In the past also, attempts have been made to make use of the magic word 'new', to initiate policies that have been discredited afterwards, e.g., the new economic policies of Rajiv Gandhi. Where the package of 'new' economic policies, introduced since July 1991, is 'new' is in its very strong emphasis on 'globalization' and 'liberalization', as opposed to a policy of self-reliance by way of import substitution. The essential features of the policy are as follows: (i) integration of the Indian economy with the world market, by reducing tariff barriers, import quota restrictions and permitting the operation of foreign companies in all spheres, more or less on equal terms with the domestic companies; and (ii) reliance on the market to carry out the function of allocation of resources, and leaving it to the prices of goods, interest rates, wage rates, foreign exchange rates, etc., to equalize demand and supply. The Government has virtually no economic role to play.

Both are dangerous for India. Such integration with the world economy can only take place on the terms dictated by the developed countries. MNC s would take full advantage of it, at a time when the world economy is undergoing a serious recession, in order to boost their exports, while it is doubtful whether Indian exports would gain. Such integration would lead to specialization based on comparative advantages; and in India's case that would mean more emphasis on agriculture and less on industrialization. The IMF conditionalities plus the Dunkel proposals would ensure that India would remain perpetually dependent on the rich countries for technology and knowledge, and thus an independent path of capitalist development would be virtually impossible for India to follow under this 'new world order'.

Further, such integration would be limited to spheres where only the rich countries would gain — unlike the free and unrestricted flow of capital and goods that they are demanding, they would not permit an equally free and unrestricted flow of manpower from India to those countries. Even though India and other poor countries enjoy a comparative advantage in terms of this factor of production — manpower — and despite a market demand for those in the rich countries of the world, such migratory movements are unlikely to be encouraged by the protagonists of globalization. Large-scale subsidies on agriculture continue to play their role in the rich countries, both in the United States and in many countries of Europe, though these are not permitted in India under IMF conditionalities. Further, no matter how devoted they are to the principle of free movement of goods, the United States would not allow Indian textile exports free access to the US market: countrywise quotas have already been fixed under the 'multifibre agreement', though such quotas are being discouraged by the IMF and the World Bank in the case of India. Similarly, US companies are up in arms against Japanese exports of cars and electronics, and are urging their Government to take punitive action against what they describe as 'dumping', and to pressurize the Government in Japan to buy more of US goods, irrespective of whether the US consumers have greater liking for Japanese products or whether the Japanese consumers are keen to buy US goods or not. Whether the market wants it or not, 'Do what I say, don't do what I do', appears to be the dictum.

The fact that the rich countries are capable of applying such double standards today is largely because of the disintegration of the socialist camp. Compare the present situation with that prevailing in the early seventies, when the Third World countries, including India, joined hands to demand a 'new international economic order'. Under this banner they sought an end to the domination of world trade by the rich countries and their MNCs, the specific demands being: an equalization of the terms of trade between industrial products and the primary products; the curbing of the domination of the multinational companies over their economies and of their various abuses; declaration of the economic sovereignty of the countries over their own resources; and a fair share of the industrial output. Today, still recovering from the trauma of the past few years, no one talks about that or the South-South dialogue any more. The post-1991 'New World Order' appears to be ruling now. The United Nations has also been reduced to the rubber stamp it was in the early years of the organization when most of Africa and Asia were under colonial rule and the socialist camp was weak. To understand India's plight now, one has to keep this qualitative transformation in the international context in mind. Globalization, on the terms of the rich countries, cannot be fought by India alone, but India being a very large country, can set examples for others to follow. In the fifties, India did play such a leadership role along with Yugoslavia and Egypt; the time has come to lead again.

In India there is now a tendency to take a romantic pre-1929 view of market-capitalism that does not exist anywhere in the world today. The unquestioned faith in the market to cure all ills cannot be justified by the experience of the world economy in the post-1929 period. The market only responds to price signals and purchasing power; the multitude of poor and unemployed, in great need but with no purchasing power, would not be recognized by the market. By a process of 'cumulative causation', the market has the tendency to make the rich richer and the poor poorer. The market-oriented approach is also, almost as a rule, a short-term approach, since the discounted present value of future earnings beyond 9-10 years is very close to zero at the present rate of interest. The alarming implications of such short time-horizon decisions for the environment cannot be ignored e.g., the Bhopal gas tragedy. The market also leads to instability, periodic crises, and can never create a condition of full employment, as the *magnum opus* of Lord Maynard Keynes clearly suggests: nor can the market take care of the needs of the future generation.

While controls and subsidies should be less, no poor country can ever industrialize without resorting to these. The governments of South Korea and Taiwan also followed strict control regimes with large public sectors until they developed sufficiently to be able to take on the rest of the world. In India's case, though on paper elaborate controls existed, in practice each control had a price and could be easily dispensed with by judicial use of money power. Here the difference is as much in terms of the quantum and the nature of control as in its implementation. It is important to note here the distinction made by Gunnar Myrdal between 'hard' and 'soft' States, the former being those where rules are meant to be implemented and the latter those where these are as a rule violated.

This brief survey of the economic developments over the past few years clearly indicates the position taken in the core of this article: that without land reforms and an active natural policy on import substitution, with a heavy reliance on human resources and the 'market', the path of independent capitalist development will continue to elude India.

The Economy in Transition

DHIRES BHATTACHARYA

I

It can truly be said of the Indian economic situation that the more it changes, the more it appears to remain the same. Since the termination of colonial rule in the country at midnight, 14 August 1947, a spate of economic policy-measures have been tried to influence the pattern of economic life in the country. But the impact of these measures on the modes of working and living of the vast majority of the country's inhabitants remains rather superficial, especially in the rural areas.

The reasons for the apparent changelessness of the Indian economy are not far to seek. Economic policies in India have been directed more at the creation of new types of industrial and commercial ventures rather than towards the resuscitation of traditionally established callings. The inevitable outcome of this has been the bypassing of large numbers of workers in the conventional occupations and the coming into prominence of a small number of specially favoured groups. Changes in living standards among these latter groups have been remarkable, but they enjoy their new-found prosperity among surroundings where deprivation and misery still abounds.

It has often been remarked that planned economic development in India has an 'urban bias'. This would be taking too generous a view of the outcome of the process, since most of the smaller towns of India have been bypassed in the development policies. Urban municipal services have been allowed to stagnate or even decline, while new industrial townships have absorbed the funds that could otherwise have been used for the upgrading of existing services. Instead of building up from already existing bases, planned development in India has most often tended to create altogether fresh bases of development, with the result that only a few areas have come to enjoy the benefits of change, whereas the rest of the country, urban as well as rural, has continued to live in an 'area of darkness'.

One indication of this almost complete obliteration of the smaller urban areas from the country's development map is to be found in the distribution of the country's urban population. Leaving aside the very large metropolitan areas, there has been a slowing down in the growth of most other urban settlements. For example, in 1951 nearly 43 per cent of the urban population of the country lived in towns with less than 50,000 inhabitants. In 1981 the proportion of the urban population in these relatively small settlements had declined to as little as 28 per cent. This was not because the larger cities were multiplying in number, but because they were able to offer at least some opportunities for a better living to those who migrated to them. In fact, in point of numbers it was the medium-sized towns, with population ranging between 10,000 and 50,000, which were growing at the fastest rate. But in spite of a growth in numbers their capacity to persuade people to stay with them was evidently declining because whereas in 1951 they accommodated 30 per cent of urban dwellers, in 1981 this class of towns provided a habitat to only about 24 per cent of the total urban population.

Of India's 557,000 villages nearly 80 per cent had come to be served with electric power by the end of 1988-9; but if the benefit of the spread of electricity is to be judged by the number of pumpsets energized with this power, the unevenness in the diffusion of benefits leaps to the eye. The available figures appear to indicate that almost 88 per cent of the energized pumpsets are localized in only 9 out of 25 States, the highly benefited States comprising about 58 per cent of the country's total geographical area and containing about 657 per cent of the total population. Per square kilometre of area, the State of Punjab has the highest concentration of electrified pumping equipment, followed by Tamil Nadu and Haryana. A similar lack of balance can be found in the distribution of fertilizers and irrigation water as well.

Economic changes are also dependent on social and cultural predispositions. Parts of the country, where traditional caste rules and clan leadership patterns are still quite dominant, have been less affected by the changes that have occurred elsewhere. This has also been due to an upsurge in the growth of population in these areas. A study of the growth of population in India between 1981 and 1991 reveals very pointedly that different parts of the country have responded quite differently to the policies pursued over the last thirty-five years or so in respect of family planning. For 1981-91 the decadal growth of population for India as a whole has been put at 23.56 per cent which was slightly below the 24.66 per cent growth rate of the previous decade. Yet there were five States where the decadal growth for 1981-91 exceeded 30 per cent, and without exception all these five States had tribal communities as their principal inhabitants. Of the eight States where the decadal growth rate was above 25 and below 30 per cent, all except one were under the dominance of caste, sect or tribal rules. The exceptional case is Maharashtra where migration may have been an important factor behind the slightly higher rate of growth during 1981-91 as compared to the previous decade.

Changes have also remained unnoticed because of lowering of the mortality rate. This has been the outcome of a substantial measure of success in the control of disease-carrying vectors, bringing down the death rate per thousand of population from 27.4 in the decade 1941-51 to an estimated 12.2 during the quinquennium 1981-6. Though the birth rate has also declined over the years, the survival rate has gone up from decade to decade, the rate in rural areas being always slightly higher than that in urban settlements.

The limited scope for migration from rural to urban areas has meant that a very large proportion of the growing workforce had to get absorbed in traditional farming and farm-related activities. Returns from an extension of such activities to relatively poor land and other natural resources being low, vast numbers of people had to face a depressed level of subsistence, with no hope of adding to the meagre income earned in traditional occupations in the village. For these people (almost one-third of the total population of the country, according to sample surveys) the quality of life has not changed in the least. In fact, due to lesser availability of employment in the face of growing numbers, general insecurity has most probably increased.

II

For the various reasons noted above, as well as administrative laxity that has often made a mockery of well-conceived policies (land reforms, for example), the perception has gained ground that in forty years, from 1951 to 1991, economic planning in India has failed to change the face of the economy. The fact, however, is that while the inner core of the economy has almost retained its previous character, the outer crust has seen momentous changes. Taking metropolitan development as an example, although there has been little increase in the number of large cities, there has been considerable vertical expansion in all such cities and, wherever possible, over-spilling of people into satellite townships has also taken place. Without such development the number of the people residing in the larger cities and towns (with more than 100,000 residents) could not have moved up from 27 million to 94 million between 1951 and 1981. During this period while the urban population as a whole went up by 150 per cent, the numbers living in the premier cities and towns of the country increased by almost 250 per cent.

Traditional sources of power and energy have become virtually obsolete in the urban locations. In their place electricity has come to be used for almost all domestic and industrial purposes. From only 6.6 billion kilowatt hours in 1950, electricity generation mounted to 236.3 b. kwh. in 1988-9, thus considerably increasing the per capita utilization of electrical energy. Inter-city air connections have almost eliminated distances. The distances flown by aircraft in domestic service increased from 21.6 million kilometres in 1950 to 61.1 m. km. in 1989, but owing to the larger capacity of the aircraft employed the number of passengers flown registered an increase of 3000 per cent. The number of passengers carried by the Indian Railways increased three-fold between 1950 and 1988, while the volume of goods transported went up three and a half times. In the case of both goods and passengers the average distance travelled increased by at least 50 per cent. More roads came to be macadamized adding to the convenience of modern wheeled traffic. In fact, between 1950-1 and 1988-9 the number of registered motor vehicles of all types increased more than fifty times, the fastest rate of increase being recorded in respect of two-and three-wheeler vehicles.

There has also been commendable development in postal communications. The number of rural post offices is now more than four times as many as in 1951, while urban posts offices have increased at least three times over. There has been an almost sevenfold increase in the length of the surface mail route covered by the Indian post offices. The number of telephone connections — a meagre 168,000 for the entire country in 1951 — has now reached a figure of nearly 5.5 million. New facilities like Subscriber Trunk Dialling and International Trunk Dialling have been introduced and are being expanded. Yet it is also to be noted that India trailed behind China, Pakistan or Sri Lanka in respect of development of telecommunications even in the late eighties.

In India, development of the basic infrastructure — power, transport and telecommunications — has been the responsibility of the Union and State Governments. In addition, beginning from 1956-7, the year when the country's Second Five Year Plan was launched, the responsibility for nurturing the heavy machinery industries was also thrown upon government agencies. The central idea was to direct investment towards the machine-building sector which alone could provide a solid foundation for rapid industrial development. In the decade 1956-66 public sector investment, mostly in the heavy and basic industries, accounted for roughly 60 per cent of the total investment of the country. This brought about a remarkable degree of diversification in the industrial structure. The importance of traditional consumer goods industries diminished in course of time, while the weightage of non-traditional manufactures, mostly machinery and industrial raw materials, increased in the index of industrial production. Comparing the weightage schemes in the years 1956 and 1980-1 one discovers that machine manufacturing was relatively almost four times as important in the latter year than in the former. At the same time the relative importance of consumer goods industries was almost halved.

This diversification of the industrial structure, however, involved a great deal of dependence on foreign sources of fuel, crude metals, some metal manufactures and many basic chemicals. The import bill escalated too fast for export earnings to catch up with it. In most of the years since 1956-7 the current account of the balance of payments has been in deficit and official capital inflows were needed to settle the external accounts. Up to the end of the financial year 1984-5 a sum of Rs16,200 crore was borrowed from diverse official sources, while non-official borrowing also went up. According to the latest available figures, total external liabilities, official and non-official, now exceed Rs 100,000 crore. About 10 per cent of this debt now constitutes the annual external servicing liability (amortization plus interest) of the country.

GENERATION OF INVESTIBLE RESOURCES, EXPECTED AND ACTUAL, IN INDIAN PUBLIC SECTOR ENTERPRISES

Period	Expected	Actual
1961-62 to 1965-66	450	395
1969-70 to 1973-74	1966	1596
1973-74 to 1977-78	2667	954
1980-81 to 1984-85	7697	5791

Note: While estimates were expressed in some base year prices, the actual figures were available in current year prices only. Since prices were rising throughout the periods concerned, the actual resource generation figures were overestimated in real terms.

The new set of industries that have come up after 1956-7 have been on the whole unsuccessful in generating either domestic investible resources or exportable commodities and services. In several of them the investment has been unproductive; in others subsidies have somehow maintained the level of output. The way expectations have been belied by investment in public sector industries can be discerned by comparing expected and actual resource generation by these industries from 1961-2 onwards.

The dominant role played by Government agencies in developmental activities, and their indifferent performance has given rise to a host of problems. First, the Government has to assume responsibility, not only for setting development in motion, but also to maintain and service the assets created in course of development. This means in effect that expenditure during any Plan period leaves in its wake the liability for a large volume of non-Plan expenditure, intended simply to keep the system going; the result very often is the under-utilization and under-maintenance of assets created during an earlier Plan period, while each new Plan aims at creating other newer assets. Secondly, in order to assert the supremacy of the public over the private sector, a plethora of regulating agencies have been brought into existence. This has also added considerably to the Government's non-Plan expenditure at all levels. Thirdly, in order to equip itself with finance to carry out all these responsibilities the Government has resorted to public borrowing on a fairly large scale; most of the time market borrowing alone has not been enough to meet the almost unending need for financial resources. Hence, resort to the Central Bank has been a frequent phenomenon, leading to large-scale monetization of the public debt. Recently in 1989-90, interest payments on past loans constituted nearly 13 per cent of all Government outlay (Union, State and Union Territories taken together). In the same year the amount of non-Plan expenditure of all types was no less than 65 per cent of the Government's total expenditure. The addition to the Government's liabilities to the public (including the Reserve Bank of India) was equivalent to nearly 35 per cent of total Government outlay. This implied that interest liability would continue to mount in course of years, roughly at the rate of 3 per cent of the total outlay.

Curbs on private investment were imposed both in order to prevent the concentration of economic power in a handful of oligopolists, and to centralize the allocation of scarce foreign exchange. Industrial houses with more than a minimum amount of assets were not allowed a free hand in choosing their most prospective fields of investment. Government approval schemes of joint ventures with foreign firms were virtually ruled out. Inflow of foreign venture capital was thus reduced to a mere trickle and, as a consequence, Indian firms tended to fall behind foreign firms in technological attainments. This reinforced the tendency to become more and more dependent on foreign sources for essential equipment and components, which was of course quite at variance with the objective of becoming self-sufficient in respect of machinery and other accessories to which mere lip service was paid in the successive Five Year Plans.

III

With Government borrowing mounting, without appreciable returns from the assets created on the basis of such borrowing, a time was bound to come when the credibility of the Government would be suspect. The basically vulnerable position of the Government came to be exposed in all its stark reality in the months following the US-Iraq War. The foreign exchange resources had to be used up in larger and larger amounts as the prices of oil soared, since it was unrealistic to hope that the demand for this vital fuel would prove to be elastic to prices. Lack of confidence in the Government's ability to cope with the external imbalance led to speedy withdrawal of deposits held by non-resident Indians in Indian banks during the summer of 1991. Drastic reforms were obviously needed to reverse the situation and to prevent a recurrence of such crises in future.

Since further import compression could take place only by incurring loss of valuable domestic output, the major burden of adjustment necessarily fell on efforts to augment exchange earnings, through exports of goods and services and other means. In exploring such means, there came about the realization that the regime of excessive controls on private investment was the biggest hindrance to recovery. There had to be an about-turn in the hitherto followed policy of according primacy to public enterprises, whatever their achievements might be, and freedom of decision-making had to be restored to private entrepreneurs if the economy was to be restructured. Hence the most recent changes in Indian public policy have been concerned with underplaying the role of the public enterprise system, except in certain special fields of action, and to unshackle the private economy by removing counter-productive regulatory measures.

The new policy measures were initiated with a devaluation of the Indian rupee in July 1991. In fact, the prevailing rate of exchange of the rupee in relation to the major currencies of the world had become unmaintainable. A more realistic rate was needed to make Indian exports competitive in the World market, though it was recognized that price competitiveness might not by itself be adequate to boost exports. The lower value of the rupee was also intended to reassure foreigners (including non-resident Indians) entrusting their funds to Indian credit agencies, that the Government would not stand in the way of their getting a proper rupee equivalent of their hard-earned money.

Under the prevailing system the major policy instrument for stimulating exports was a system of cash subsidies. This not only aggravated the domestic budgetary imbalance, but also invited retaliation. Accordingly, a new system was devised which would automatically enable exporters to realize high domestic sale values through the market itself. Initially, exporters were allotted import licences, generally at about 30 per cent of their foreign exchange earnings, which they could utilize themselves or sell to importers at a premium. The draft on free foreign exchange for imports was thus reduced, while exporters were lured by the higher profits into larger sales of exportables. Subsequently the freely exchangeable components of export earnings and other foreign remittances was raised to 60 per cent, only 40 per cent being handed over to official agencies at a lower official rate. Several other concessions were simultaneously introduced for export-related imports.

Far-reaching changes were also introduced in the domestic economic arrangements. A virtual halt was called to the expansion of the public sector, and units in that sector which could be described as chronically sick were referred to a rehabilitating agency — the Board of Industrial Finance and Reconstruction — for proper disposal. Restrictions so long imposed on large industrial houses were removed at one stroke. The requirement of a licence for establishing new industrial units or expanding and diversifying existing ones, was virtually withdrawn. Clearance of proposals for foreign collaboration with Indian industrial enterprises was made almost automatic; only the foreign exchange angle remained to be scrutinized. Financial reforms were also taken in hand to assist exporters and industrial entrepreneurs. All in all, the changes favoured a return to private decision-making processes, the planning exercises being limited to special activities like extension of infrastructural facilities and special employment-generation schemes servicing the needs of local pools of unemployed and under-employed workers.

Opinion on the probable consequences of these important changes remains sharply divided. On the one hand, it is being argued that the freedom to invest, restored by these changes, is likely to be abused. A few industrialists will be in a position to establish their hegemony over the economic system, but their smaller counterparts and the consumers in general are certain to lose. Moreover, as new mechanized and semi-automatic processes come to be introduced, employment in the modern industrial sector must shrink; rehabilitation of the displaced workers will not be an easy task and may generate social unrest as workers' unions resist the changes. On the other hand, these fears are dismissed as baseless by those who speak of the possibilities of almost endless future growth being opened up after a period of senseless stagnation under a regime of controls and regulations. It is only through growth, they argue, that the unemployed and the poverty-stricken multitude can be drawn into the economic mainstream. Subsidies and special relief measures can serve their purpose only for a short while and only so long as Governments are able to divert funds to these ends from more productive activities. In the longer run, they point out, there is no alternative to growth-based redistribution for which the ground has to be cleared first.

Economic theory by itself can do little to provide credibility to either of the two opposing views referred to above. Various non-economic variables, such as the spirit of enterprise displayed, the equation between the workers and their employers, and the overall supervision exercised by the Government for preventing aberrations, will ultimately determine the course of future events. But it deserves notice that along with countries of the erstwhile Socialist bloc, India (which had also formally embraced Socialism, though it was practised in a milder form) is also on the brink of great changes, the full consequences of which cannot be determined at the present moment. For at least the next decade considerable uncertainties will prevail, policies being alternately softened and tightened as their outcomes turn out to be acceptable or otherwise. The end of the experiment will very likely give us a different compound of private and public participation in economic management than what we have been familiar with so far.

Taking stock of the situation three years after the Government of India introduced a series of 'momentous' changes in basic economic policy, one cannot fail to notice that the momentum of the changes has been anything but considerable. Such changes as have occurred belong to the periphery rather than the core of the system. Poverty and unemployment continue unabated, the rate of inflation has become more severe, and the process of growth has fallen short of the rate attained during the ten years immediately preceding the introduction of the package of reforms. In calling a halt to the growth of the public sector, the policy-makers have failed to take adequate note of the fact that over the last forty years public and private enterprises have learnt to grow in tandem, albeit with a very slight (often negligible) improvement in technical efficiency. Their bonds cannot be cut asunder so easily and so fast. Restructuring of the industrial economy, in particular, is facing new problems of working out a *modus vivendi* between the old-style public sector units and the upcoming private units keen on introducing the latest products and techniques.

Perhaps it is this incompatibility between the old dispensation and the new that has been partially responsible for the very slow growth of output in the modern industrial sector during the last few years. The average rate of growth in the manufacturing sector during the years 1980-1 to 1990-1 was 7.6 per cent per year. But in 1991-2 this rate became negative and in the following year there was only a modest 1.2 per cent expansion. Figures available for the period April to October 1993 reveal that output grew at a slower rate than in the corresponding period in 1992. Indications are also available that the depression was deepest in the capital goods manufacturing industries on which future expansion depends.

With sharply reduced responsibility for the promotion of new industrial units the Government is now in a position to divert more funds for social improvement and special employment-generation projects. The needs in these respects are however so vast that moderate increases in financial allocation for these purposes can have only a marginal impact. The really crucial elements determining the future course of the economy are, first, the rate of growth of fixed capital formation and secondly, the employment-elasticity of the investment in such capital. On neither count is the latest available information reassuring. Fixed capital formation in the country has undeniably slowed down compared to the years preceding 1990-1 and the new types of capital are probably even less employment-generating than those installed in the previous decades. It looks as if the transition to a more liberal economic system will be marked by an initial downturn, the recovery from which will depend largely on how much the rest of the world responds to India's calls for help in activating her economic potential.

Planning in a Liberalized Economy

PRANAB MUKHERJEE

Planning for economic development was conceived as the principal instrument for bringing about material progress and a transformation of society. India's economic scene has undergone significant changes under the impact of four decades of developmental effort, as a result of which it has been possible to pool together a fair measure of skilled manpower and ample entrepreneurial resources. It has also been possible to build up a sound and resilient agricultural economy coupled with a diversified industrial structure. The growth in GDP and consumption levels crawled for almost two decades and thereafter we could achieve and sustain a growth rate of 5.5 per cent per annum during the eighties. The upshot was a perceptible increase in consumption levels and reduction in poverty. Science and technology took noticeable leaps.

During the eighties agriculture forged ahead at more than three per cent per annum. This was made possible by development of the agricultural infrastructure and supply of better inputs combined with pricing and public food procurement policies. It was a good performance. Given the purchasing power and its distribution, the country enjoys self-sufficiency in food although that dates back to the late seventies. It may be argued that this has provided a greater range of policy options in international affairs.

Another thing that has to be highlighted is the development in the infrastructure during the eighties — coal, oil, power and transport, and capital goods made long strides to put the economy in a position to claim around eight per cent growth in industrial output. What brightened the scene was the reduction in imbalances, through planning in these infrastructural sectors, which had dislocated production up to the mid-seventies.

During this period, the Indian economy showed a capacity to absorb external shocks. While the world economy in general faced an economic recession, India bypassed it through timely investments in oil and gas development. In the recent 5-7 years, a growth rate of seven per cent in oil consumption has been sustained and international market fluctuations in the wake of the Gulf War could not drastically dislocate our domestic production processes.

The sound macroeconomic position achieved in the early eighties enabled a liberalization of the 'domestic' economy pursued in the Seventh Plan period. It has not been often acknowledged that the macroeconomic stability achieved over the years of planning has enabled India to honour its international debts. But that has been possible only with massive help from international financial institutions. Initially, India had a good credit rating and these bodies were willing to lend money on low conditionalities. But as the country's need for loans increased, conditionalities became more stringent. Accordingly, in the changed international situation, the country's capability to honour debts has to be strengthened still further.

The pace of capital accumulation, which includes human capital accumulation, leads to a 'learning effect'. Its allocation to infrastructure projects helped in reducing their gestation period. In procuring equipment from abroad — for oil, power, coal, etc — right questions were asked and the imported capital equipment could be procured on reasonable terms. The decline in incremental capital-output ratio in the economy during the eighties can be explained to a considerable extent in terms of improvement in resource-efficiency in the capital intensive projects. For instance, the capability of power stations based on domestic equipment is about the same as of the ones with imported equipment.

On the other hand, the negative side is presented by the sluggish growth in employment, significantly lower than the increase in labour force. Underemployment persists and poverty, although reduced, is still high. Literacy has improved but there are still miles to go. Provision of social services — health, housing and education — has been far below acceptable levels. The position is critical in many districts and pockets. Adequate resources for providing a social security net to the population are not in sight. The rate of growth of population too has not gone down because of the inadequacy in the social infrastructure.

There is a feeling that progress has been much slower than expected. The process has been wasteful. The public sector has done many things which are essential but in the name of doing the essential it took in hand much that was inessential. In the name of protecting the poor, it has frittered away much of the resources on those who are not so poor. Being screened off from the market system for long, the public sector has become oblivious of the standards of performance and norms of efficiency. The attempt to insulate the economy from foreign competition served us well initially in building up an industrial base but in course of time left us way behind global economic and technological advances. This has also engendered infectious inefficiency in the private sector. Our growth has been slow and costly.

In recent years we have experienced acute problems in financing development, particularly in the public sector. Because of internal fiscal deficits and foreign debts, which have been accumulating with an increasing burden of debt servicing over the years, the foreign exchange situation faced a crisis a few years ago. Two things needed to be done. The crisis had to be handled, and a reorientation of the strategy of development had to be effected.

A spate of economic reforms have resulted of late. These are devaluation; allocation of foreign exchange through partial convertibility of the rupee, rather than through direct exchange control; curtailing allocation through licensing, paring criteria in industrial licensing only to the single criterion of self-financing of foreign currency needs; almost free choice to foreign capital in selecting the area of investment; and a willingness to accept the consequences of these measures on organized sector employment. These policy thrusts aim at giving the market a greater role in determining choices and allocating resources : and that calls for a reorientation of planning.

THE ROLE OF PLANNING

Those who put planning and the market as alternative systems, perhaps, do not realize that the two could best be seen as, and made, complementary to each other. Prof. Sukhamoy Chakravorty in his famous book *Development Planning* makes a fine distinction between the 'market mechanism as an efficiency promoting device' and 'the market as a process which has to be valued in itself' and which is based on 'consumer sovereignty'. As far as the market as an efficiency-promoting device is concerned, it is very consistent with planning. It is the second view of the market as a value in itself which gives paramount importance to consumer sovereignty, that may not have a place in our socio-economic context as yet. Here a few points may be made. A regulatory apparatus tends to set up signals which are damaging from a social and economic point of view: thus pointed control leads to smuggling; control of drinking habits leads to bootlegging and worse; and fixation of prices and rationing result in blackmarketing and hoarding. The planning process should be keyed to take account of such distortions, otherwise they may overwhelm objectives as in eastern Europe. In India, the political apparatus is sensitive, so some corrective steps can be taken before a systemic background is threatened. Thus 'indicative' planning acquires importance in a sophisticated economy.

Centralized planning of the type practised in socialist economies never existed in India. The share of the public sector in investment has been less than 60 per cent and its share in production (value added) is just about 25 per cent. Thus the market has determined prices and allocations in a major segment of our economy throughout the planned development in India.

Planning interventions have varied from sector to sector and from time to time. Agriculture has been almost entirely developed by the private sector, even though a large part of the agricultural infrastructure was provided by the public sector. Such infrastructure includes irrigation, fertilizers and credit. Industries such as those of power and petroleum, where large investments are needed, have been developed in the public sector. Initially coal was in the control of the private sector but as demand soared, the public sector assumed the role of developing it. In industries such as fertilizers, petrochemicals, steel, and services like road transport, the role of the private sector has declined, as the public sector has accumulated the potential to develop these areas. This process will expand and extend to other industries and services in the Eighth Plan.

The removal of large disparities in development between regions requires a flow of resources across them. Experience has shown that market forces have not achieved this objective in adequate measure. The planning process has to manage this flow of resources for accelerated removal of regional disparities both in agriculture and in industry. At present the Plan does provide for special area programmes such as the Hill Areas Plan, Tribal Areas Plan and other schemes for backward areas. While these programmes aim at creating a basic infrastructure, the backlog of development is large and considerable effort is to be made to integrate such regions into the mainstream of economic activity in the country. For the welfare of the poor, who have little asset-endowments to benefit from the natural growth of economic activities, planning is very essential. This planning process develops the social infrastructure which embraces education, health, and social mobility, and reduces disparities between social groups with the help of public support. It is freed from the bureaucratic control that is exercised over trade and industry and is directed towards social engineering. The political energies of a democratic setup are far better spent on such activities which do not involve mindless file work.

In the Directional Paper to the Eighth Five Year Plan it has been stated that planning in India will be mainly indicative. Planning in our country in fact has in large measure been indicative. An outstanding example is planning for agriculture. It is a sector where millions of farmers take independent decisions. However, when we decided to achieve self-sufficiency in food, we indicated certain production targets and set about creating conditions in which the farmers could be induced by their own self-interest to take decisions which would help attainment of targets. What all this involved was making available a high yielding variety of seeds through research which was wholly Government-sponsored, creation of an infrastructure of irrigation systems, roads and markets, and providing a package guaranteeing minimum prices or procurement prices. To this had to be added an adequate supply of essential inputs like energy and fertilizers along with extension of education.

Recent changes have not diminished the vital need for planning. It is essential in order to tackle the gap in the physical and social infrastructures which still exist in our country. It is also necessary to bring about a certain degree of rationalization in a system which has become wasteful in some areas. The thrust of recent economic policies is towards this goal.

In future we will enlarge the area of indicative planning. We have to sharpen our tools and develop our methods still further. We have to find out which policies are effective, which investments are catalytic in nature and which institutions work.

An important pre-condition for indicative planning is to have a strategic vision of the future, i.e. the direction in which we wish society to proceed. That vision may include an articulation of our aspirations in respect of human development and material progress. That vision would also broadly indicate the path before us. The Planning Commission has a perspective planning division which looks at the long-term perspective of the economy as a whole. Five Year Plans fit into that perspective.

Let us now look at the strengths and weaknesses of the new role the Planning Commission is expected to perform and the need to reorient it. In the entire governmental system, the Planning Commission is the only institution which, by virtue of its position of having to deal with all aspects of the economy on behalf of the Centre as well as of the States, is capable of taking an integrated and interrelated view of the total economic system. Other departments of the Government deal with single sectors and therefore have a sectoral view of developmental priorities. The States particularly, have their own regional view of national priorities. The Planning Commission alone can have an overall view, and has developed capabilities in terms of data systems and models over the years to take such a view. It can cope with responsibilities which cannot be handled efficiently or equitably by market forces. These include public-good aspects and such externalities. It is true that as the market takes on more and more of allocative functions, there will no longer be need for a very detailed tie-up between demand and production of various commodities. However, given the nature and stage of our development, the building up of a hard-core infrastructure will still remain largely the responsibility of the Government for some time. Then the long-term views of population, natural resources like land, water, and the environment, will necessarily remain the concern of the Government. If one has to plan even for these limited areas, one has to take an integrated view of the entire economy. The macroeconomic constraints of parameters like savings, balance of payments, and fiscal balances will need to be monitored and will have a major role in delineating the path of development. The Planning Commission will, therefore, have to keep on performing this function even though the market will be taking an ever expanding responsibility in the area of production and distribution of goods and services.

In the new context, however, there is a need to reorient the thrust the Planning Commission exercises. The emphasis on detailed accounts of commodity balances and inter-industry consistency has to be reduced as the market starts playing an increasingly larger role in allocative systems. There will be greater need for monitoring the major trends in the economy, short-and medium-term forecasting and studying the price responses. Models to perform these functions will have to be constructed.

EMPHASIS IN THE EIGHTH PLAN

The Directional Paper to the Eighth Plan makes a plea for a re-examination and reorientation of the role of the Government in this context. It envisions the role of the Government as being one of creating an environment which encourages and builds up people's initiatives rather than making them dependent on the Government and which liberates the forces of growth and modernization. The State has to play more of a facilitating role. In more specific terms, it envisages that: (a) people's initiative and participation should be made a key element in the development process; (b) the role of the Government should be to facilitate the process of people's involvement in developmental activity by creating the right type of institutional infrastructure, particularly in the rural areas; and (c) the public sector plan will have to become very selective, making investments only in those areas where investment is of an infrastructural nature or for human development facilitating growth and development as a whole. Investment in the first case will be essentially for preservation and augmentation of the basic resources of the country like land, forest, water, and science and technology. Where investment is for human resource development, it will be for improving health and education, and for protecting the weak and the poor whom market forces tend to overlook.

The public sector will continue to perform an important role in the coming years. However, certain critical weaknesses that are now apparent will have to be addressed. The public sector, as envisaged at the dawn of planning in India, was to have contributed to the growth and development of the nation by providing more reinvestible resources. This has not happened in the manner it should have. Many PSUs make substantial losses and have become a continuing drain on the exchequer, absorbing resources which are desperately needed to achieve other developmental goals. Apart from the fact that the fiscal situation does not permit accumulation of unsustainable losses, there is also the fact that many PSUs which are in the red do not serve the purpose which they were set up for.

There may have been very good reasons in the past for the public sector to take initiatives in industrial areas where the private sector either would not enter or would hesitate to do so. This may not be the case today, and the restructuring of the public sector would essentially entail vacating such areas for private sector initiatives in the coming years. Recent Government policies have already envisaged vacating large areas for private sector initiatives and this process of restructuring and reform needs to be carried further.

In the new situation, the public sector may not be expected to maintain its conventional share in investment. However, even after accepting the most selective role for the public sector, there is the 'requirement' side to be looked at. Since the projects have a long gestation period in the infrastructure sector, investments in the immediate future determine the longer term growth possibilities for the economy. In the two recent annual plans (1990/2), the investment in oil, coal, power, railways, and irrigation was confined generally to the already approved ongoing projects, and new projects for future needs, in general, could not be provided for. Infrastructural bottlenecks can lead to a situation of the kind that prevailed in the mid-seventies when shortages of cement, steel, coal and power were severe, and the annual rates of inflation had crossed 20 per cent. Private sector initiative can reduce the need for public sector investments. But some lag between policy changes and actual investment is inevitable. Though the power sector has been opened up for the private sector, it will take some time before adequate response materializes.

It is pertinent in this context also to point out that subsidies need to be rationalized so that they are not frittered away on sections of society for which they were not originally meant, and also to see that they do not serve as inducements to inefficiency. Therefore subsidies to the public distribution system, subsidies for education and health services, subsidies for essential services like transport and energy, and subsidies for fertilizers need to be earmarked for the targeted sections of society. Subsidies also lead to waste of resources most of the time. A very relevant example is the subsidization of water rates in agriculture. It leads to waste of water, encourages flood irrigation and sometimes causes water-logging and salinity problems doing permanent damage to the soil which is difficult to repair. Irrigation subsidies also lead to sub-optimal cropping patterns.

As mentioned above, the long-term objective is to switch over from the 'coercive' to the 'indicative' mode of planning in every sector. Once a 'macroeconomic' framework is devised, planning should proceed on a selective sector-specific basis. While going in for a particular type of planning intervention, we will have to be much more discerning than in the past with regard to each sector. A particular kind of intervention adopted for a sector should necessarily be reviewed every 2-3 years. One of the predicaments of planning has been the tendency to avoid reviewing and reforming the institutions and modes of intervention. It has to be borne in mind that experience has been gained in the method and working of different institutions in different parts of the country and the emphasis should be on bringing the know-how of one part of the country fruitfully to bear upon another. State politicians and bureaucrats regularly visit the Planning Commission. Apart from the written documents which are circulated, a considerable amount of information is disseminated through the agency of the Planning Commission. The role will have to be much larger. The Planning Commission has to play a leading role in the indoctrination of the process to be emulated by other planning institutions.

The danger to our ecological balance stems not so much from industrialization as from the fast-growing population dependent on the finite resources of land, water, flora and fauna. The time horizon of the market is too circumscribed to focus on these issues. Here the flow of benefits has a much longer time dimension than what the market can evaluate and take into cognizance. Sectors which cater to the health, educational, and other basic needs of society also belong to this category: here the planning intervention has necessarily to be 'coercive'.

It may be mentioned here that, without fanfare, India has been associated with the cause of environmental promotion for a very long time. The concern came into prominence at the Earth Summit held in Rio de Janeiro in 1992. Incidentally, the *World Development Report* 1992 published by the World Bank focuses on India's contribution to tackling environmental problems.

There are certain special characteristics of the sectors mentioned in the preceding paragraph that need to be emphasized. Firstly, the quantification of benefits is difficult to achieve unlike in the industrial projects. Secondly, the flow of tangible benefits to the economy begins after a much longer time than in conventional projects. Thirdly, the organizations and institutions that implement the schemes and programmes here are not and cannot by nature be as well-structured as industrial establishments. Fourthly, the 'concerns' tend to be reflected in the 'outlays' demanded rather than in the 'effectiveness of the expenditure'. The measurement of benefits loses its objectivity if the administrative agency that implements the programmes also measures the benefits. The special characteristics of the social welfare sectors require that a sound 'extra-administrative agency' is made responsible for measurement of benefits and evaluation of results.

The distinction, between groups which will benefit from public expenditure and those which will not, has to be rigidly adhered to, if the limited resources are to show up tangible benefits. Further, the implementation apparatus for such programmes should move much closer to the location. Control through a highly centralized apparatus results in multiplicity of programmes which try to reach the same location. The variety of centrally-sponsored schemes is an example of this. These should give way to 'integrated area packages'.

We have come to a turning point in our perceptions on the ways of achieving a complete social and economic transformation. The Eighth Five Year Plan pays more attention to democratization at the grassroots level, to involvement of people in the process of change and to the building up of institutions appropriate to the process of change and adequate for the needs of the people. We hope to smoothen the path of this transition and accelerate the processes of growth and development.

Perspectives and Recent Policy Changes

ARUN GHOSH

The purpose of a development policy is to hasten changes in the system of production and distribution so as to improve the standard of living of the people. The process through which such changes have been sought in all countries (other than a few like China) is 'growth': growth in agricultural and industrial output, and higher productivity of labour, brought about through faster accumulation of capital, leading to a more rapid pace of development, which is then supposedly shared, through increased employment at steadily higher levels of productivity, by the entire workforce. The First Five Year Plan, commencing from April 1951, presented the problem succinctly in the following words:

> The urge to economic and social change under present conditions comes from the fact of poverty and of inequalities in income, wealth and opportunity. The elimination of poverty cannot obviously be achieved merely by redistributing existing wealth. Nor can a programme aiming only at raising production remove existing inequalities. Only a simultaneous advance along both lines can create conditions in which the community can put forth its best efforts for promoting development.

There can be no doubt that the achievements of the Indian economy during the past five decades have been quite significant. One of the most remarkable features has been the success of the Indian economy to accommodate more than a double of the population, from around 360 million in 1951 to 844 million in 1991 — the population growth rate being of the order of 2.2 per cent per annum – and yet bring about an average real growth rate of the economy, significantly higher than witnessed at any time in the history of the Indian economy. Food production has more than trebled, from around 52 million tonnes in 1950 (admittedly a bad year) to 171 million tonnes in 1990 (also, not one of the peak production years). Industrial production has also recorded an average growth rate of between 5 and 6 per cent per annum (the exact percentage not being cited here because of periodic changes in the index of industrial production, and doubts expressed in certain quarters as to the comparability of these data over time). During the decade of the eighties, *real* GDP recorded a growth rate of 5.5 per cent per annum. By all statistical parameters of growth the Indian economy may be stated to have 'arrived' at the threshold of self-sustaining growth, with a gross domestic savings rate of between 21 and 22 per cent of the GDP (which had touched 23 per cent in one year), an economy capable of producing most of the capital goods and key intermediates required for continuing growth of output, and a developed economy in a position to feed its teeming millions.

And yet, India today is facing a crisis of no mean order, a crisis which is multifaceted, a crisis in its socio-economic framework, a crisis which has spilled over to the political arena and threatens to tear the country apart.

As far back as 1964, Professor P. C. Mahalanobis – himself the architect of the Second Five Year Plan and the model of development which was pursued thereafter, though in fits and starts – sounded the warning signal when he presented a report on the distribution of income in the country. Mahalanobis came to the painful conclusion that the distribution of income had certainly not improved, if not worsened, in India since Independence and specifically, that the concentration of economic power in private hands had increased to an unwarranted extent.

Coming to more recent times, as per the Census of 1991, forty-eight per cent of the Indian population of the age of 7 years and more was illiterate, as of 1 March 1991. The extent of illiteracy was much higher in some States, significantly greater in the rural areas, the more so among women, and particularly disturbing among the depressed sections of society, especially among the Scheduled Caste and Scheduled Tribes people.

The National Sample Survey (conducted over July 1987-June 1988) has already brought to light certain other disturbing developments. As per a detailed analysis undertaken by Profs. B. S. Minhas, L. R. Jain and S. D. Tendulkar, in 1987-8, as many as 43 per cent of the people of India were below what may be described as the 'poverty line', the percentage being 45 for the rural areas and 39 for the urban areas; and the incidence of poverty was particularly acute in some States (*Economic & Political Weekly*, Bombay; 6-13 July 1991).

Successive rounds of the National Sample Survey have also revealed the rapidly falling elasticity of new employment in relation to GDP growth. Between 1972-3 and 1977-8, overall employment in the country grew by 2.82 per cent per annum. Between 1977-8 and 1983, the growth of employment declined to 2.22 per cent per annum. Between 1983 and 1987-8, employment growth declined further to 1.55 per cent per annum. Thus, in the eighties, when the growth of the GDP attained the highest rate achieved since Independence, employment growth was slower than the growth rate of the population and of the workforce, leading to an increase in visible, overt unemployment, in addition to the substantial degree of low productive employment and under-employment (especially for farm labour) characteristic of a developing economy.

Greater access to education being one of the major stimulants to development, Prof. B. S. Minhas has analysed the 1986-7 National Sample Survey data, relating to the population of age six and above. In the rural areas, 43 per cent of males and 69 per cent of females (and, in the aggregate, 57 per cent of all persons) had never enrolled in a school. In the urban areas, 17 per cent of males and 36 per cent of females (and in the aggregate, 26 per cent of all persons) had never enrolled in a school. The reasons were partly economic compulsions but very largely because adequate (or satisfactory) school facilities were just not available.

The broad point is that over the forty years of planned development, output did increase significantly, but the distribution of the output has been highly skewed; the benefits of growth did not reach a large segment of the population. Development has also been concentrated only in a few regions; in many others, there has in fact been a profound degradation of the environment, including, in some areas, a decline in the moisture retention capacity of the soil because of large-scale felling of trees, the formation of gullies during the monsoon and the washing away of valuable top soil, and the increasing desertification of large areas of the country. As a result, consumption — even of food — has not increased (on a per capita basis) in many areas in the country, and in some areas has recorded a decline. The per capita consumption of pulses, the only major source of protein for the poorer sections, has recorded a steep decline in the country as a whole, and so has the consumption of millets and inferior cereals, the staple diet of the poor in the rural areas.

To put it briefly, not only has the distribution of the benefits of past growth been highly lopsided, even the resources available for development have deteriorated for large sections of the population.

It is in the above context that, of late, the need has been felt for a change in the very strategy of development planning. The approach to the Eighth Five Year Plan, approved by the National Development Council in October 1990, sums up these concerns and chalks out a new strategy. However, that document became infructuous because of the turn of events; and as it happens, the country was overwhelmed in late 1990 early 1991 by a balance-of-payments crisis of an unprecedented order, coupled with an increasing fiscal (or domestic macroeconomic) imbalance, which in fact, has been handed down for very nearly a decade. The response, since mid-1991, has been a series of dramatic economic policy changes which have set out an entirely new paradigm of development policy. It is important, in this context, to examine some of the roots of the economic crisis which overwhelmed the economy in late 1990 early 1991, the policy responses thereto, and some of the dangers that the Indian economy faces in the coming years.

THE NATURE OF THE PRESENT CRISIS

Some of the emerging problems have already been mentioned, while projecting the experience of planned development in India in the past four decades. The economic crisis which overtook India during July 1990-June 1991 was essentially twofold: (i)a persistent macroeconomic imbalance in the economy arising principally from budgetary deficits of the Central Government throughout the eighties, accentuated in the years 1986-90; and (ii) an increasing deficit in the balance of payments which was sought, again from

1986-7 onwards, to be made good by short-term borrowing of various kinds from abroad (including 1-3 year deposits in foreign currency from non-residents at high rates of interest). The budgetary deficit of the Central Government on revenue account, which was Rs 1,715 crore in 1980-1, shot up to Rs 5,888 crore in 1985-6, Rs 7,776 crore in 1986-7, Rs 9,137 crore in 1987-8, Rs 10,515 crore in 1988-9, Rs 11,912 crore in 1989-90 and Rs 17,585 crore in 1990-1 (revised estimates). As percentage of the GDP, the revenue deficit of the Government of India averaged 1.1 per cent during the Sixth Plan period (1980-1 to 1984-5), and 2.6 per cent in the Seventh Plan period (1985-6 to 1989-90).

The budgetary deficit of the Government sector (including the State governments) created a macroeconomic imbalance in the economy which spilled over into the balance of payments. The payment problem was further exacerbated by the liberalization of imports and of foreign collaboration, leading to the setting up of a growing number of import-intensive industries designed to produce consumer goods in increasing demand from the elitist groups in the country. The gap between gross domestic investment and gross domestic saving (both as percentages of the Gross Domestic Product at market price) increased from 1.4 per cent in 1984-5 to 2.1 per cent in 1985-6, 2.2 per cent in 1986-7, 2.3 per cent in 1987-8, 2.8 per cent in 1988-9, 2.5 per cent in 1989-90 and again 2.8 per cent in 1990-1. (*Economic Survey* 1991-2; *Currency & Finance Report* 1990-1). The steadily increasing gap between domestic investment and domestic saving was reflected in a deficit in the balance of payments not matched by net receipts of external aid, and was made up by short-term borrowing from abroad. As per details given by the Reserve Bank of India, short-term borrowing (made up of commercial borrowings, non-resident deposits, and borrowings of one year or lower) were more than 60 per cent of capital inflows in 1987-8, 69 per cent in 1988-9, and more than 70 per cent in 1989-90. The net inflow of external assistance accounted for less than 40 per cent of capital receipts in 1987-8, and less than 30 per cent in each of the years 1988-9 and 1989-90. (RBI Bulletin, July 1991).

Indeed, it was this dependence on relatively short maturity-borrowing from abroad which created a problem — by way of capital flight from India — because of loss of international confidence after the Gulf War, and the sudden increase in payments for oil imports thereafter. Between October 1990 and October 1991, over a period of thirteen months, non-resident deposits in Foreign Currency (or FCNR deposits as they are briefly designated) alone declined by a net amount of $ 1.84 billion. There also arose problems in the rescheduling of commercial borrowings, especially of shorter maturities. Borrowing from the IMF under the gold tranche, under the CCEF, and under the first credit tranche — to the tune of approximately Rs 4,500 crore between October 1990 and March 1991, which incidentally approximated the higher payments for oil imports following the Gulf War — were not adequate to meet the problem arising from the flight of short-term capital from India, leading to the danger of default in meeting all payment liabilities, in the summer of 1991.

It is perhaps well to emphasize here that: (a) since 1987-8 in particular, exports from India had been increasing remarkably, through the increase in imports; and (b) that the crisis in the balance of payments, particularly in the first half of 1991, had in a sense been building up over a long period, though it was precipitated by a flight of short-term capital and not by any immediate problem of current payments, especially after the severe import restrictions imposed during that period.

Exports from India in 1986-7 increased by about 15 per cent in rupee terms, and 10 per cent in dollar terms. In 1987-8, exports increased by nearly 24 per cent in rupee terms, and 21 per cent in dollar terms. In 1988-9, exports increased by about 26 per cent in rupee terms, and close to 23 per cent in dollar terms. In 1989-90, exports increased by 36 per cent in rupee terms, and more than 18 per cent in dollar terms. Such buoyancy in exports had not been experienced ever before in the past. Even in 1990-1, a year of crisis, exports increased by 17.5 per cent in rupee terms and 9 per cent in dollar terms. However, the gap remained wide; and more particularly, much of the gap was made up by short-term borrowing, and not long-term development assistance, as in the past i.e., until the seventies.

THE STRATEGY ADOPTED — JULY 1991 ONWARDS

The response of the new Government, elected to office in June 1991, to the immediate balance of payments crisis has been threefold: (i) an immediate devaluation of the rupee (by some 20 per cent), mainly to stem the flight of capital, accompanied by a stringent policy of import restriction, export promotion (largely through transfer of most imports of raw materials and components to financing under saleable Exim scrips granted to all exporters), and an application to the IMF for a loan of $ 2.26 billion (in the upper tranche) to meet the balance of payments problem; (ii) a macro-economic stabilization programme, seeking to bring down the 'fiscal deficit' of the Government of India (i.e. total Government expenditure *minus* Government's current revenues) from 8.4 per cent of the GDP in 1990-1 to 6.5 per cent of the GDP in 1991-2; and (iii) a programme of structural reform, with a view to making Indian industry more efficient and internationally competitive.

There have been some further policy changes since, in regard to industrial policy and reform of the financial sector; and more particularly to certain far-reaching changes introduced in the budget for fiscal 1992-3, presented in February 1992. These changes include, *inter alia:* (a) partial convertibility of the rupee (exporters and remitters of service incomes getting 40 per cent of their foreign exchange earnings' remittances converted at the official rate and the balance at a free market rate of exchange, and commercial imports as well as all capital transactions — other than with the IMF — being at the free market exchange rate, with specified imports being allowed at the official exchange rate); (b) the scrapping of the scheme of granting Exim scrips to exporters; (c) significant liberalization of the procedure for the inflow of foreign capital; (d) continuing emphasis on macro-stabilization (planning to bring down the fiscal deficit of the Government of India to 5 per cent of the GDP in 1992-3); (e) continuing focus on structural reform by significant liberalization of import policy, accompanied by discussions with the IMF World Bank for a large loan (of an amount varying from $5 to 7 billion) under the Extended Fund Facility of the IMF.

The short-term 'adjustment' problem does not concern us here, but in so far as the short term is the precursor of the long term, some of the policies adopted need to be explained in a little more detail, with a view to examining the long-term implications of these policies.

THE BUDGET AND THE CREDIT POLICY

The most important points arising from the budget are twofold. First, the focus has been on the reduction of the 'fiscal deficit' (not on the revenue deficit, which, in 1991-2 remained very large), and its reduction was sought to be achieved partly by the sale of a part of the equity of selected, profitable (and highly undervalued) public sector enterprises, and partly by drastic cuts in budgetary support for both social infrastructure (like education, health, employment-generation programmes) and for the build-up of economic infrastructure. Secondly, there was in fact a reduction in direct tax rates (both for the income tax and the wealth tax), with a view presumably to create more space for private enterprise. Thus, the attention of the Central Government has been focused not on revenue-raising but on cutting the Government's investment expenditure. Total taxes (both of the Centre and the States) in 1990-1 were no more than about 17 per cent of the GDP, and direct taxes again were no more than about 17 per cent of total tax receipts (RBI: *Currency & Finance Report* 1990-91). At the same time, the needs of the stabilization programme have led to the adoption of a stringent credit policy, raising interest rates to previously unheard of levels, which *inter alia* had led to the stagnation of industrial production (the index of industrial production in April–December 1991 having actually recorded a decline — with promise of a continuing decline in industrial output during 1992 — after an average growth rate of around 8 per cent per annum in the eighties).

PROBLEMS OF ADJUSTMENT AND TRANSITION

Several fundamental problems have arisen with the adjustment and structural reform programme initiated by the Government since July 1991. The first relates to the *speed of adjustment*. The drastic cuts in Government expenditure have ushered in a deep recession in the industrial economy at a time when long-term employment generation was in any case lagging behind population growth. Secondly, the stabilization programme endeavours to attain macro-economic stability primarily through changes on the demand side — through a curtailment of Government expenditure on social programmes and through credit curbs (for production) — which is typically an IMF prescription and works reasonably well in industrially developed economies, but not so well in a developing economy where there are basic supply constraints requiring large infrastructural investments.

There is herein a basic conflict between the short-term needs of adjustment and the medium-term objectives of expanding output and increasing efficiency of production. The severe restraints on domestic production coupled with the continuing devaluation of the exchange rate of the rupee have not helped the balance of payments situation either: the year 1991-2 having recorded a 1.9 per cent decline in the dollar value of exports from India (exports in rupee terms having recorded a small increase of less than 6 per cent).

The unwillingness to raise direct taxes and, in particular, the reluctance to tackle the problem of growing tax-evaded (or black) incomes, have continued to fuel inflation, the wholesale price index having recorded an increase of some 13 per cent in 1991-2, the increase in consumer prices being significantly greater. Even in regard to the programme of structural reform of industry, several problems arise. The cutback on infrastructural investment could create serious bottlenecks in basic infrastructure, particularly in regard to power and transport. The sudden withdrawal of support from public sector enterprises — without any systematic attempt at basic reform, such as the distancing of enterprise from Government administration — is likely to lead to general demoralization and a setback to morale among the workers and staff. In fact, there are already signs of an agitational response to the Government's unformulated proposals concerning an 'exit policy' for industry. The pattern of the adjustment programme, as already indicated, is such that the expenditure-cuts fall almost entirely on investment (and on plan outlay), while the burden of tax (and price) increases on the poor. For example, the percentage of capital expenditure with respect to total expenditure is seen to decline from 32.8 per cent the Seventh Plan period to 26.1 per cent in 1991-2, and further, to 24.8 per cent in 1992-3. Plan outlay as a percentage of total expenditure shows a similar decline from 37 per cent during the Seventh Plan period to 29 per cent in both 1991-2 and 1992-3. The cuts in expenditure on social sectors implies that the burden of adjustments is likely to fall almost entirely on the poorer sections of the population. At the same time, the reduction in the rates of direct taxation and the absence of any attempt to track down and tax black money, together with relaxation in import policy, are clearly oriented to benefit the richer sections of the population. This is reflected in the unprecedented boom in the stock market where the *average* price : earnings ratio has, as of writing, reached the absurd figure of 50 : 1 (implying a 2 per cent return on investment). The phenomenal increase in share prices, together with an increase of between 20 and 30 per cent in food prices reflects who precisely is shouldering the burden of 'adjustment'.

Is the process of adjustment undertaken from July 1991 sustainable? If the process of planned development from 1951 to 1991 has led to increasing inequality in the distribution of income in the country, and in fact in the continuation of illiteracy, ignorance, and poverty among a large section of the population, the programme of 'globalization' of the Indian economy, of linking it up integrally with the world market is likely to lead only to further marginalization of increasing sections of the population. The social tensions that such a process is likely to create would not be conducive to an ambient atmosphere which is essential for increasing efficiency and productivity.

The policy of 'partial convertibility' is unlikely to boost exports, and in fact, there will be no self-balancing mechanism between imports and exports, such as may have been possible, through the linking of commercial imports to Exim scrips (or to a part of designated export earnings). The 'partial convertibility' of the rupee — which is in effect a complex multiple exchange rate system — is unlikely to achieve the intended result of encouraging exports, and could only lead to the under-invoicing of exports, while the over-invoicing of imports (for siphoning off part of the investment cost of projects by the promoters) would not get checked. In fact, the liberalization of imports at this juncture is likely to continue to strain the balance of payments.

The programme of stabilization — in particular, the credit curbs introduced — have already led to a decline in industrial production. The impact on small-scale industries has been particularly adverse, and many of them have closed down, accentuating the problem of unemployment. The partial withdrawal of the State from the area of social infrastructure, in particular, education, health and welfare, is likely, in the face of increasing elitist consumerism (widely publicized through the television network), to accentuate social unrest and tension.

There is already evidence of deepening fissures and the fragmentation of society. The increasing centralization of authority and the reliance on forces of 'internal security' to preserve law and order, the dissension and divisive forces that have sprung up in different regions of the country, the growth of a parallel economy, the increasing lack of gainful employment opportunities, are all likely to intensify the criminalization of society already apparent in some parts of the country. The programme of stabilization and adjustments — introduced at the behest of the IMF/World Bank — is unlikely to produce any long-term benefits on the Indian economy.

CONCLUDING REMARKS

The basic solution of the endemic problem of poverty and low productivity in India calls for a multifaceted approach. To begin with, there is need for giving the maximum attention to rural development generally and to employment generation in particular. This can be achieved economically only through a process of decentralization and of local area planning, with some form of employment guarantee. This would give higher incomes and at the same time help to regenerate degraded soil through land reshaping, terracing, planting of trees and grasses, and other diverse measures to improve the moisture retention capacity of the soil in dry areas and to prevent water-logging in low-lying, flood-prone areas. Increased incomes are likely to stimulate the demand for industrial products, for which the focus should be on the production of mass consumption goods. The long-term improvement of labour productivity would result partly from 'local area planning' and partly through better education, health, and higher skill formation.

The balance of payments problem can be easily surmounted, partly through a sustained effort to equate domestic saving with domestic investment and partly through a self-balancing mechanism in regard to payments for commercial imports and a part of export earnings. Macroeconomic adjustments will occur if people are suitably enthused, if the burden of adjustment is appropriately shared (primarily by the rich), if unproductive and wasteful government expenditure on defence, and outlays on very long gestation megaprojects are deferred.

The present programme of adjustment and structural reform contains within it elements of self-destruction. If large external borrowings are resorted to in order to sustain the adjustment programme, the burden of external debt would increase and the country would fall into a debt trap from which it would be difficult to extricate it. More importantly, the present reliance on increased inflow of short-term external capital (e.g. by way of non-resident deposits) can lead to the same difficult situation as was created in the first half of 1991. A large loan from the IMF at this juncture can only engender complacency, at the risk of future default by India on its external payment obligations.

POSTSCRIPT

A lot of changes in policy have been rung in since this essay was written. There is, however, no reason, as of mid-April, 1993, to change any of the conclusions reached earlier. Further data available as of now only buttress the conclusions reached earlier.

At the outset, it should be stated that the objectives of the 'stabilization' policy are twofold: (i) to bring about a better macroeconomic balance within the economy, and (ii) to bring about a better balance in external payments. There is also a third objective of the policy of 'economic reform', namely, to increase the efficiency and competitiveness of the production process in the country.

As far as the macroeconomic balance within the economy is concerned, it can be stated that the policies adopted since July 1991 have, by and large, succeeded. Inflation, as of the end of March 1993, is down to 6.7 per cent on a point-to-point basis over March 1992. However, the exact opposite appears to be the case in the matter of external payments. As per the *Economic Survey* for 1992-3 (published by the Ministry of Finance), the deficit on the current account of external payments is likely o be $ 7 billion in 1992-3, as compared to $ 2.8 billion in 1991-2. As per the *Economic Survey*, the total external debt of India, as of 30 September 1992, was $ 7.1. billion. To this figure, one has to add $ 10.3 billion of rouble credit (in terms of the recent agreement reached with Boris Yeltsin) and $ 2.2 billion of military debts to GCA countries, making India's external debt total to $ 83.6 billion. According to the *Reserve Bank of India Bulletin*

Supplement for November 1992, the repayment schedule for India's external debt over the next three years is as follows:

	US Dollars (million)		
	1993-4	1994-5	1995-6
1. Principal (of which, FCNR/NRER deposits)	5447 (1026)	6014 (357)	5616 (–)
2. Interest	3516	3226	2900
Total	8961	9240	8516

To the above figures, one would have to add approximately $ 770 million in respect of repayment of the rouble credit, and an unknown sum in respect of military debt to GCA countries. And, if the repayment of the principal in respect of FCNR/NRER deposits is avoided by renewal of these deposits, one would have to add further interest payments. The balance of payments deficit on current account (of $ 7 billion in 1992-93) includes the payment of interest, but excludes the repayment of principal. If note is taken of the debt contracted after December 1991, the repayment figure would certainly increase further.

During April-December 1992, exports (in dollar terms) increased by only 3.4 per cent over exports in April-December 1991, against an increase in imports by 16.5 per cent (also in dollar terms). Thus, the policies adopted since July 1991 may be stated to have signally failed to tackle the balance of payments problem.

While fiscal and monetary policies have succeeded in bringing about better macroeconomic balance within the economy, this has been achieved not through increased tax revenues from the affluent sections of the population, nor by curtailing current expenditures by the Government, but by (a) a severe curtailment of investment in infrastructure, (b) a reduction in *net* Plan transfers to State Governments, and (c) the sale of part of the equity of profitable public enterprises (and taking credit for this capital account receipt on to the revenue account). For 1993-4, while the overall budgetary deficit has been brought down sharply, the revenue deficit continues to remain as high as in the past. To that extent, the macroeconomic balance has been achieved by sacrificing future growth: as a result, there has been an increase rather than a reduction in unemployment, the focus has shifted away from social sector outlays on education and health care and there is no attempt to redress the absolute poverty in many regions of the country.

Finally, as a result of the combination of several factors, while the 'efficiency' of a few manufacturing units has improved, there is no evidence of any increase in overall factor productivity in the economy as a whole. If anything, on an aggregate basis, average factor productivity may well be declining. Part of the reason may be weak morale, particularly in the public sector, which provides infrastructure and key intermediates. The reduction in investment is likely to show up in future by way of a tottering infrastructure. At the same time, while the liberalization process has happily eliminated many bureaucratic controls, there is also evidence of increasing centralization and concentration of power. The steady increase in the expenditure on paramilitary forces in order to maintain law and order is symptomatic of the gradual distancing of the ruling elite from the masses, which is not a sign of increasing efficiency of economic agents (of production) in the country. The unmistakable shift in focus away from long-term development cannot hope to improve overall efficiency and productivity which is a major objective of the economic reform package.

Industrial Reforms in India: Retrospect, Prospect and Challenges

AMIT MITRA

INTRODUCTION

This paper on industrial policy in India is presented in two parts. Part I deals with the roots of the 'license-quota-control' *raj*. Over fifty years, ideas on political economy of development were played out and some of them crystallized into policies. Now that these policies are being rethought, it is very important for us to trace the evolution of these ideas in history. Only then can we fully understand the gravity of our current task.

Part II of this paper deals with the industrial reforms taking place since 24 July 1991. Though some references have been made to the roots of reforms (given in Part I), Part II is almost self-contained. It tries to conduct a critical and constructive evaluation of the current industrial policies and attempts to draw lessons for the future. These two parts will enable the reader to get a full measure of the depth of our challenge today.

PART I

INTERNATIONAL MILIEU OF FUNDAMENTAL CHANGES: RELEVANCE FOR INDIA'S REFORMS

We are living through historic times. The hiatus over the validity and efficacy of alternative systems of political economy has reached a new plateau. The strident intellectual and material thrust of communism has ended with the collapse of the Soviet Union.[1]

Further, confusion on the far left has reached a peak with China adopting the 'free market system' without dismantling its totalitarian one-party (Communist?) State. These rapid and startling developments have sounded the death knell to Central Planning and have called to question the very capacity of any government to conduct Macro-Five-Year-Planning, even in its indicative form.[2] In fact, the 'market system' has undoubtedly emerged as the credible alternative, sending its reverberations to the once socialistically inclined Third World.

New Debate in the West among Pro-market Schools

However, a new debate has also opened up in the West on the *degree* to which markets alone can be entrusted to allocate resources satisfactorily. The moderate pro-market advocates are locked in an exciting intellectual debate with the *philosophic right* which supports free market institutions not merely from the narrow economic perspective but from the broader political perspective of human freedom.[3] On the other hand, the 'new institutionalists' are revolting against mainstream economics, demanding that socio-economic institutions and political processes be included in understanding economic change and development.[4]

Social Democrats in Europe: Rethinking Policy?

The democratic socialists in Europe have not escaped the intellectual ferment of our times. They are faced with the growing prob-

1 Some like Francis Fukuyama have even gone to the extent of calling this the 'End of History', where the excitement of the clash of ideologies has come to an end. See, Francis Fukuyama, 'End of History', *The National Interest*, Summer, 1989.

2 See a comprehensive treatment of this issue by Don Lavoie in *National Economic Planning: What is Left?*, Cato Institute, 1985.

3 The works of Friedrich Hayek and Ludwig von Mises (thinkers of the philosophic right) neglected for decades, are being re-read widely even in the East European nations today side by side with the works of popular mainstream economists like Milton Friedman and Paul Samuelson.

4 See, Douglas North 'Institutions, Transactions Costs & Economic Growth', *Economic Inquiry*, July, 1987, vol. XXV, no. 3 and 'Transactions Costs, Institutions, & Economic Performance', Occasional Paper No. 30, International Centre for Economic Growth, 1992. Also see, Mancur Olson, *The Rise and Decline of Nations: Economic Growth, Stagflation and Social Rigidities*, 1982.

lems of the Scandinavian economic experiment, once a shining example for the world. The confusion in its ranks was evident at the 18th Congress of the Socialist International at Stockholm in June 1989. In its Declaration of Principles it was loathed to say that '... experience has shown that while nationalization in some circumstances may be necessary, it is not by itself a sovereign remedy for social ills'. Yet, in the very next paragraph it reaffirmed its traditional faith in nationalization by stating that 'the democratic socialist movement continues to advocate both socialization and public property within the framework of a mixed economy'.[5] The world-wide negative outcomes from Public Sector Enterprises (PSEs) seem to have merely touched but not transformed the views of social democrats, yet. However, in practice, the dividing lines on economic policy between the right and the Centre in (say) Germany seems to have blurred with much greater sensitivity towards markets and private enterprise on the part of German social democrats, and greater willingness on the part of the Christian Democrats (the right) to involve the government in rebuilding the erstwhile East Germany.

Technological Revolution: The Implications

In other words, we are living through a phase of human history that is relentlessly breaking down past belief structures. It is also producing quantum jumps in technology, simultaneously. The latter is opening up vistas hitherto unseen by human eyes. Technological change is also forging deeper connections within and between nations through revolutionary modes of information processing, *simply multiplying the free market's ability to facilitate business transactions.* In this milieu of change, the forces of economic globalization are irrevocable. However, the real challenge of our times lies in shaping these megaforces to benefit the largest number *within nations* and include *all* nations of the world in this new sweep of history. All this makes obvious the fact that the current economic reforms taking place in India *cannot be viewed in isolation* of these momentous changes.

5 Socialist International's Declaration of Principles, published in India by B B S Chaudhury at National House, Bombay, p. 15, paras 59 & 60 respectively.

International Underpinning of our Model of Development Collapses

Needless to say that direct connections between the current international reordering and economic policy thinking in India are well established. After all, India's planning models were deeply rooted in those of the Soviet Union: a nation that has suddenly ceased to exist and a system of planning that has been abandoned by its own people. Furthermore, many of the Third World nations who were our partners in learning from the Soviet experience have long since jumped ship and are now hitting new highs of economic liberalization and growth.

On the other hand, India's 'government centre' industrial policies of the past were also rooted in the theoretical works of *western economists* who were quick to support market based private enterprise models for their *own countries,* but opposed the same for Third World economies. They were of the view that markets in developing nations suffered from acute rigidities and were incapable of allocating resources effectively. Furthermore, entrepreneurship was supposedly lacking and the capacity to save (and invest)was absent.[6] Therefore, they recommended the Governments of Third World nations to take the lead and orchestrate industrial development through planning, controls, state ownership and corrective price distortions: a policy position now critically called the *'dirigiste dogma'*.[7] Yet, today, those very same Western economists, some of whom had helped India to set up its Five Year Plans and build its labyrinthine State control mechanisms, have recanted their past position.

The mounting evidence against the efficacy of the 'State-centered' model has compelled them to rethinking. In other words, the very *underpinning* of the policy path pursued by India has been steadily undermined by the forces of history and by decades of mounting counter evidence. This has made the radical liberalization of India's economy totally inevitable. However, what may have hastened the pace, leading to the significant reforms begun in 1991, is the balance

6 See, a non-technical essay by this author substantiating this view in *Seminar* April, 1991, no. 380, pp. 13-17.

7 See, Deepak Lal, *The Poverty of 'Development Economics'*, pp. 5-16, for a pointed discussion of the notion of *dirigiste dogma*

of payments crises and the plummeting of hard currency reserves to a mere one month of payments for imports.

UNDERSTANDING THE ROOTS OF INDUSTRIAL POLICY IN INDIA: THE DEEP FOUNDATIONS OF THE STATE-CENTRED APPROACH

In order to fully understand the importance of the industrial policy changes initiated on 24 July 1991, one must first unearth the roots of industrial policies in modern India. One must then plot the evolutionary path of these policies, taking note of the inner struggles, to either transcend these policies or deepen the control mechanisms. Obviously, the recent initiative is an integral part of this long history and, therefore, the need to look back and reflect for a moment.

Avadi Session only Furthered the Process of Policy-making

It is popularly believed that the Avadi Session of the Congress Party, held in January 1955, was the watershed for industrial policy in India. It defined a course that has remained with us in principle through these decades. There is no doubt that the Avadi Session was the most powerful force in laying down the politico-economic basis for crystallizing the industrial policy in post-independence India. It unambiguously declared that '... to further the objectives stated in the Preamble and Directive Principles of State Policy of the Constitution of India, planning should take place with a view to the establishment of a socialistic pattern of society, where the principle means of production are under social ownership or control...'[8]. specify that the State will 'initiate and operate large-scale schemes providing services such as power, transport, etc.' These words aptly summed up the tenor of State ownership for the next three decades!

Furthermore, the State was to '...have overall control of resources ...'. It would also 'check and prevent *evils* of anarchic industrial development ...'. The semantic and the rhetoric of Avadi is important to note too. Words such as 'social ownership', 'control', the '*evils* of anarchic (albeit, free market) industrial development' have rung through these decades in text books, newspapers and popular

8 All India Congress Committee, *Resolutions on Economic Policy, Programme and Allied Matters* (1924-1967), p. 86

speeches. In a way, these words have taken on lives of their own in the Indian social milieu. Of course, Avadi's firm directives were soon institutionalized and operationalized in the Industrial Policy Resolution of 1956 — a document that features in every Industrial Policy Resolution ever since.

Digging Deeper into History: The National Planning Committee's Clear Directives

Digging through the past a little deeper, one discovers that the Industrial Policy Resolution of 1948 in fact contained almost all of the 'socialistic' ingredients of the post-Avadi Resolution of 1956, without of course, taking the rhetorical high ground.[9] Obviously, the 1948 Resolution had political sanction for keeping three industries under the monopoly of the State. It must have drawn strength from some existing sources to have kept six industries under the 'exclusive' charge of the State for establishing new undertakings. These included such crucial areas as iron and steel, manufacture of the telephone and telegraph, and mineral oils. Furthermore, eighteen industries were slated for Central regulation and control, mirroring the later policies of 'industrial licensing' formalized in 1951 — all this, well before the Avadi Session! Obviously, there were earlier sources from which the 1948 Resolution took guidance.

The answer to this comes from the protracted deliberations of the National Planning Committee (NPC) appointed by Subhas Chandra Bose and chaired by Jawaharlal Nehru himself. A staggering number of 77 meetings of the NPC were held, stretching from 17 December 1938 to 9 July 1946, and Pandit Nehru recorded his presence in 76 of the 77 meetings.

In three crucial Notes and Memoranda to the Members of the NPC, Nehru forcefully defined the premises on which the NPC would draw in formulating its recommendations. In these Memoranda, he repeatedly drew the attention of the Members to the Karachi Resolution of the Congress (1931), and specifically to these words: 'The State shall own or control key industries and services, mineral resources, railways, waterways, shipping and other means

9 Note that the Parliament had accepted the national objective of creating a 'socialistic pattern of society' barely a month before the Avadi Session, in December 1954. Therefore, the socialistic incline of the 1948 Resolution must have drawn strength from earlier political processes.

of public transport.'[10] He emphasized that "The general indication of Congress policy [quoted above] is of vital importance and applies not only to public utilities but to large-scale industries and enterprises which are likely to be of monopolistic character'. Then he went on to add that '*A legitimate extension* of this principle would be to apply it to *all* large-scale enterprises'. Nehru concluded the Memorandum in the form of a directive to the Members of the Planning Committee which read: 'It is clear that *our Plan must proceed on this basis* and even if the State does not own such enterprises, it must regulate and control them in the public interest.'[11] The message to the members of NPC was loud and clear.

In the final Memorandum, Pandit Nehru went into the specifics of the State-centered model. Defence industries would be owned and controlled by the State. Public utilities would be owned by 'some organ of the State'. Regarding key industries, the majority were of the opinion that they should also be State-owned, though a substantial minority considered that State control would be sufficient.'It was made clear, however, that any control of such industries must be a *rigid* one.' For non-key industries '... it was made clear that the very nature of planning required control in some measure. What this measure should be, might vary with the industry in question.'

Giving Shape to the Principle of State Control: 1948-1969

By the 1930s the political process had already outlined, and even specified to a degree, the Industrial Policy for an independent India. While there were leaders who stood on the political right of Pandit Nehru, there were also leaders within the Congress who stood to his left, making Nehru's position widely acceptable. This legacy carried us through the 1948 Resolution to the Industries (Development and Regulation) Act of 1951, and then to the definitive Industrial Policy Resolution of 1956. The control mechanisms in these Resolutions were complemented by the Foreign Exchange Regulation Act, the Import and Export (Controls) Act, the Capital Issues (continuance of Controls) Act, all in 1947. These were exten-

10 'Chairman's Note on Congress Policy', December 21 1938, in K. T. Shah, *Report: National Planning Committee*, Bombay, 1949.

11 See, for example, K. T. Shah, op.cit. p. 41.

sions of the Defence of India Act and were never to leave us. Finally, came the Monopolies and Restrictive Trade Practices Act in 1969, which simply bottled a major segment of industrial energy. Once tied together, our economic system represented the most regulated economy in the world, outside of the erstwhile communist regimes. There is no doubt that neither Pandit Nehru nor Subhas Chandra Bose, both instrumental in crystallizing some of the original ideas on controls, would have approved of the current state of affairs. They would have been the first to rethink and recast their ideas based on praxis.

The Role and Perceptions of Industrialists during the Formative Years: A Lesson for Today

Much has been made of the presence of industrialists in forums like the National Planning Committee, suggesting their acquiescence to the State-centered model. However, the evidence is somewhat open to interpretation. In the case of the NPC, it has not been recognized so far that the key representative of the industry, Mr Purshottamdas Thakurdas, dropped out of the meetings after the Third Session in May 1940.[12] G. D. Birla, who was not a Member, declined the invitation to chair a crucial Sub-committee. It has also not been recognized that A. D. Shroff, who continued to attend the meetings as a member, repeatedly recorded dissenting notes against the policy recommendations. In fact, such was the seriousness of Shroff's dissenting notes, that Pandit Nehru himself read out to the Committee his dissenting note on the Interim Report of Public Finance.[13] On the other hand, it is also a fact that the industrialists approved of Pandit Nehru's crucial 'socialistic' Memoranda cited above and had earlier agreed to a resolution on *licensing* that was the precursor to what came in 1951.[14] More disturbing is the eulogy by the authors of the Bombay Plan (produced by industrialists) to the 'economic achievements' of the Soviet Union — supposedly a five-fold growth over twelve years! It is rather strange that the industrialists had not come across works such as the *Assignment in Utopia* by Eugene Lyons which had successfully broken away leaders like

12 See, for example, K. T. Shah, op.cit. p.23.

13 See K. T. Shah, p. 185.

14 See K. T. Shah, p. 107.

Jayprakash Narayan from Soviet type Marxism-Leninism by then.[15] Furthermore, the Bombay Plan went so far as to accept under special circumstances a situation where '...practically every aspect of economic life will have to be so *rigorously controlled by the Government* that individual liberty and freedom of enterprise will suffer a temporary eclipse' [p. 55]. Such unusual concessions suggest a lack of ideological clarity on the part of the industrialists.

One could, of course, rationalize this by suggesting that the dominant socialistic belief structures and ethos of the age were responsible for these lapses and confusion. On the other hand, critiques have argued that government controls were advantageous to industrialists, particularly those who were already in positions of power in the market place. With their established positions and networks within the government, they could use the control mechanisms to keep competition at bay. Whatever may be the real explanation for a confused or at least a softer position towards the State-centered model, it is clear from history that the *collective and evolved thinking of industrialists today* may be of crucial importance to the rethinking under way.

PART II

THE EXACT STRUCTURE OF CONTROLS AND THE *MODUS OPERANDI* OF REFORMS IN 1991

The Government of Narasimha Rao obviously inherited an entire structure of industrial ownership and controls, a sense of which we have already had in tracing the roots of industrial policies through history. If reforms are to be meaningful today, it is this inherited structure that would have to be *partly dismantled* and suitably *restructured*. Therefore, it is important for us to understand the specifics of the inherited structure and to simultaneously cognate the changes made to this structure since July 1991. Such an exercise will enrich our grasp of the *modus operandi* of the new policies.

15 Narayan had read this between March 24 and April 2, 1938, well before the Bombay Plan was written. See Minoo Masani, J.P.: *Mission Partly Accomplished*, 1977, pp. 9-10.

Foreign Firms: Discretionary Domain to a New Transparency

From the very beginning, industrial policy-makers viewed Indian firms differently from the foreign-owned firms. The *entry* as well as the *expansion* of foreign firms were kept entirely in the *discretionary* domain of the Government, and only since 24 July 1991 has a reasonably *transparent* set of rules, along with a separate list of industries been developed for the entry of foreign owned companies. This major discretionary power of the past, and that too on a *case-by-case* basis, understandably bred significant corruption and obstructed legitimate and useful entry: no wonder that China received as much as $6 billion in foreign direct investment in 1991 while India received a meagre sum of $100 million. While there could be other reasons for this as well, there is little doubt that the main culprit was our deeply entrenched regulatory system.

Indian-owned Industries: Some Radical Reforms and Some Halfway Houses

Indian-owned industries, on the other hand, faced a more articulated structure of controls right from 1948, as discussed earlier. But the controls were also hemmed in by *discretionary powers in practice*, and plagued by the malady of the *case-by-case* treatment.

(a) Reserved Industries for State Ownership: Change in Principle

The policies towards Indian firms began with a list of industries which were out of bounds for them, reserved exclusively for the State — a category that is as old as the Policy Resolution of 1948, and remains firm in the Policy Resolution of 1991. Of course, the principle for including industries in this category underwent a radical change for the *first time* in 1991. The old principle of reserving 'core industries' for the Government *was abandoned.* Industries like iron and steel and ship building which had remained in this list since 1948, were finally struck off the list. The new list contained industries which dealt directly with the area of national security, or were of strategic interest to the country. Unfortunately, this new policy cannot now go back in time and reverse the acute problems of failing public sector enterprises (PSEs) created largely due to the 'core sector' reservation argument. What should ultimately be done with these loss-making PSEs is quite another issue.

(b) Compulsory Licensing: Unfinished Reforms

The second premise of the industrial policy was to *restrict,* rather than ban *entry* of private firms into certain industries. The arguments in favour of doing this date back to Pandit Nehru's Memoranda to the Members of the NPC. The logic was that the State had to control the *direction* and the *mix* of major investments in the economy in order to *rationally plan* development. After all, inputs and outputs had to be matched based on the nation's priorities ('objective function'). Furthermore, the State had to keep under control the income distribution in the country. Additionally, monopoly power had to be curbed. Therefore, compulsory licensing of the *entry* for private firms into major industries (not reserved already for the State) would be necessary.

Despite these seemingly balanced arguments, reality played total havoc with the licensing structure as well. By the early 1970s, rampant discretion and *case-by-case* basis had polluted the licensing scene, and the veneer of rational planning had worn thin. 'Rent seeking' sharks in the Government and 'rent offering' businesses made a mockery of the original arguments. And all this came to be well substantiated as early as 1970 in the seminal work of Jagdish Bhagwati and Padma Desai,[16] supported by the Reports of a host of Government Committees appointed between 1964 and 1969. Yet, nothing really changed despite a national awareness, till 24 July 1991. Then too, 'licensing' was not abolished — 18 industries continue to languish in the licensing net to this day.

(c) Controlling Large Businesses with an Iron Fist : Unambiguously Reformed

Even after experiencing the difficulties of 'licensing', the government felt the need to add more controls in 1969, this time targeting India's large business houses. The size of industrial holdings of these businesses, often interconnected through equity shares and management controls (partly set up to escape from the web of Government regulations), bothered policy makers deeply. The possibility of private firms wielding monopoly power and the supposed negative effects on the distribution of income, loomed large in their minds,

16 See Jagdish Bhagwati & Padma Desai, *India Planning for Industrialization,* London (Oxford University Press), 1970.

not to mention political clout of large businesses in a rent seeking environment.

Therefore, the Government slapped the Monopolies and Restrictive Trade Practices Act on all firms with assets above a certain size (raised to Rs 1 billion by 1991). Such firms could enter only the licensed sector and that too on a *case-by-case* basis. *Discretion* had entered the scene by the front door. Furthermore, these MRTP firms were not allowed to expand their current facilities (fixed assets) 'substantially', without seeking sanctions. Unfortunately, the policy-makers were not concerned about the advantages of the 'economy of *scale*' and what it meant for cost reductions and export competitiveness.

Fortuitously, the policy declaration of 1991 finally removed this asset limit clause for large business from the MRTP Act altogether. Even more, in a rare case of humility, the MRTP Act in investment decisions of large companies became deleterious in its effects on Indian industrial growth[17] — a fact that was privately admitted for a decade. Now, the MRTP Act could concentrate on checking *real* monopoly power in lines similar to those of the Anti-Trust legislation in developed nations.

(d) Reservations from the Bottom Up: Too Hot to Touch?

Finally, *reservations* were put in place from the bottom up for the 'small scale sector' for over 800 classified industries in the country. The 1991 reforms kept these reservations *intact* with a total of 807 such industries in the list! Understandably, with such a vast area of industries left to the 'small scale', the 'big' is bound to pretend to be 'small' (through methods already known to industries) and surreptitiously enter this sector. It must be a regulators' nightmare to monitor such a vast and complicated terrain. Therefore, a struggle between the regulators and the regulated is bound to erupt on this terrain in the near future, now that the regulators would be less 'preoccupied' with other regulations.

Two Typical Cases of Predatory Intervention: Reformed Finally

Two other types of regulatory structures will demonstrate the depth to which control mechanisms attempted to penetrate the decision-making process of private business:

17 Point number 37, *The New Industrial Policy*, July 24, 1991.

(i) In order to force the pace of *indigenization,* particularly in the electronic and engineering industries, the Government introduced the Phased Manufacturing Programme. It not only intervened deeply in the workings and allocation decisions of industries, but even the enforcement of the complicated regulations was done entirely on a *case-by-case* basis! After much resentment and wasted time and energy, the Government now humbly admits that this regulation was a '...major irritant that a large number of firms have felt in terms of discretionary power and Government's interference in business decision' [*sic*].[18] *Having recognized this, the Government abolished this regulation in its 1991 Resolution.*

(ii) Similarly, in order to wield power over private decision-making, the public sector financial institutions (the key lenders to private businesses) imposed a mandatory *'convertibility clause'*. This provided the option to the institutional lender of converting a loan into equity holding in the borrower's company. *In other words, the threat of a take-over was built into the loan contract!* Though the threat was rarely carried out, it was used for gaining leverage over private firms. The reforms of 1991 have thankfully stopped this predatory practice altogether.

The Ultimate Implications of the Control Structure

This brief sojourn through the regulatory structure reveals that aside from a case of *over-regulation,* the industrial policies suffered from a *lack of transparency.* Discretion *(case-by-case!)* was at the heart of almost every regulation, providing the politicians and the bureaucrats golden opportunities to 'rent seek'. Therefore, corruption which permeates our society today, is clearly rooted in this industrial regulatory structure. Furthermore, a variety of inefficiencies (understandably) plagued the production system. It was simply more profitable to chase and acquire a license than to work hard for increasing efficiency and reducing costs. And the *opportunity cost* of the regulatory system was simply staggering. This partly explains the extremely high-cost nature of India's industries. Add to this the losses from the PSEs and the failures of the infrastructure (owned exclusively by the Government), and you have the recipe for a near disaster.

18 *Economic Survey*, 1991-92, Pt. II, p. 81.

Our analysis would be incomplete without stating that there is an *alternative view* on the subject of controls which claims that each of the regulations invariably carried with it *escape clauses.*[19] These clauses were effectively used by private enterprise in league with the politicians and bureaucrats, escaping from much of the regulations. I. G. Patel, an advocate of this view concludes pointedly: '...in this uneasy coalition [between capitalism and socialism in India], irrespective of who presides officially, the strident voice is undoubtedly that of the capitalists' majority.'[20] Perhaps, Manmohan Singh's recent conjecture comes closer to the truth: '... a sort of functionless capitalism seems to have flourished under the protection of a professedly socialist State.'[21] I would rather invoke the idea of 'new-mercantilism' so aptly brought out in a seminal work by Hernando De Soto.[22]

However, Patel, Singh and this author would all be in agreement that distortions, inefficiencies, corruption and lost opportunities were, in fact, a byproduct of the past regulatory mechanisms.

A CRITICAL EVALUATION OF THE INDUSTRIAL REFORMS SINCE 24 JULY 1991: ACHIEVEMENTS, SHORTFALLS & SERIOUS PROBLEMS

There is no doubt that the greatest achievement of the industrial and other reform measures taken since July, 1991, is the shift in the mind-frame and philosophy of governance. Undoubtedly the Government is making a bold and *concrete* effort to shift out of the State-centered model of development and move towards a market-centered model. It is also attempting to shift from the inward looking (import-substituting) approach to the outward looking export-promoting and globally-integrating outlook. At long last, the rhetoric of the Government is not vastly different from that of policy measures flowing from it.

19 For example, an escape clause for *Schedule B* of the supposedly most State-centered Industrial Policy Resolution of 1956 was '...but in which private enterprise will also be expected to supplement the effort of the State'.

20 I. G. Patel, Silver Jubilee Lecture delivered at the Institute of Economic Growth, New Delhi, 14 May 1984.

21 See *Economic Times,* October 1, 1992

22 See, Hernando De Soto, *The Other Path,* pp. 201-30.

The original mindset of dirigisme discussed earlier ('control the commanding heights of the economy through the confiscation of privately-generated surpluses and channel them through non-market criteria to force development') has, at least for the time being, come to rest. Even if the past ethos has not been broken yet it has been pushed on to the defensive. This watershed in political economy, though not a sufficient condition, is absolutely necessary for stepping deeper into the next phase of substantive reforms.

However, in terms of dismantling the gigantic and inter-woven body of controls, the reforms have only touched the tip of the iceberg. While measures taken in specific fields appear to be giant steps, a closer scrutiny reveals that they are insufficient in most cases. In fact, the Policy Resolution of July 1991, has already been overtaken by the natural momentum of reforms, revealing the inadequacy of the initial measures. Furthermore, the regulatory structure at the grass roots level remains almost as murky as before, threatening to *jeopardize concrete responses* from industry. 'Wait and see while you keep one foot in the door' is the common refrain among both foreign and domestic entrepreneurs. Why this is the case today will be brought out through an examination of the inner inadequacies and deeper problems of the current reform.

The Problem of Reforms at the Top and a Maze of Regulations and Inspections at the Bottom

There are two distinct levels at which problems exist and need to be identified. The first is the macro level of *'core industrial policies'* and the second is the crucial 'ground level' where a different set of obstructions operate.

(a) The Problems of Core Level Industries Policies

Core level policies consist of the Industrial Policy Resolution of 1991 and changes that have followed since then. Here, it is clear that:

(i) The list of reserved industries for the public sector has already begun to *fray and pale.* For example, the oil sector included in this list has slipped out, *ipso facto*, with the opening up of the discovered oil fields,both small and big, to the private sector.[23] Furthermore, it

23 See, Amit Mitra, 'Scuttling the Nation's Future', *Business India*, 11-24 May 1992, p. 49 for some details and also see, 'Striving to Influence' *Business India*, July 20-August 2, 1992, for a summary of attempts to influence a policy change in the oil sector.

may be just a matter of time before the Government decides to reinvolve the private sector in the area of coal. The rampant inefficiency in nationalized coal fields is reaching a boiling point. Given the inseparable connection between coal and power generation, further reform will have to follow. Also, 'corporatization' of the railways (which carry coal to power houses) is inevitable, if the reforms are to maintain their momentum. It is not commonly known that the massive production units of the railways, like the wheel and axle factory in Bangalore, do not even produce a balance sheet for public scrutiny, since they are not 'corporate' entities like the rest of the 244 PSEs. These units hide behind the veil of 'departmental undertakings', concealing their real performance profiles from the public. Therefore, 'corporatization' is inevitable and this will lead to deeper private involvement in rail transport, finally skirting the reservation list once more.[24] In other words, the reserved list for the public sector will have to be cut down even further to meet the challenge of reforms.

(ii) Similarly, the list of industries needing licences compulsorily is *badly flawed.* Ghosts from the past are still haunting this list. For example, the old prejudice against the so called '*items of elitist consumption*' unexpectedly features in the new Policy Resolution, resulting in the inclusion of such potentially dynamic industries as motor cars, *all* consumer durables, and 'entertainment electronics' in the 'compulsory licensing' list![25] *This flies in the face of the very spirit of the reforms which seek to rely heavily on market signals for the allocation of resources. It also puts us back into the vicious old cycle of 'controls, rent seeking, inefficiencies, corruption'.* In fact, the maintainence of such a list serves no real purpose in the new milieu, least of all the inclusion of many dynamic industries.

(iii) An area that was simply brushed aside for reform is the reserved list for 'small scale' industries. It may sound rather impressive to say that this sector produces Rs 1553.4 billion and employs 12.4 million people.[26] But the truth is that a large number of medium and large

24 See, Amit Mitra, 'Time for Transparency', *Business India,* April 3-16, 1992 for the reasoning behind corporatization.

25 See, Annexure II in the Industrial Policy Resolution issued by the Government on July 24 1991.

26 *Economic Survey,* 1991-2, p 87.

firms understandably masquerade as 'small', drawing handsomely on the attractive fiscal and other incentives. In fact, even multinationals have been charged in the past for pretending to be 'small'. Therefore, the reforms should have addressed this issue head on, and offered *a new scale-integrative scheme for industrial production* rather than simply *avoiding* reforms in this area. Note that East and South-East Asian nations had taken an integrative approach between the small, medium and the large, and are now reaping the benefits from this approach. Couldn't we have learned from them on this issue?

(iv) Perhaps, the greatest attention has been paid to the promotion of foreign direct investment and technology transfer in the 1991 Resolution. After years of dawdling, finally, 51 per cent equity holding has been allowed for a clearly specified set of industries, and hiring foreign technicians can be done without seeking the Government's permission. The effect of this *transparency* is already visible in the number of foreign collaboration *applications* approved between August 1991-September 1992. [See Appendix for details.] However, even now, there is confusion over the jurisdiction of the sanctioning authorities — 'should one go to the high powered Foreign Investments Promotion Board (FIPB) for approval [dealing with 'mega'(?) projects] or should one go to the Secretariat for Industrial Approvals (SIA), or to the Reserve Bank of India (RBI)?' Such confusion demands a *single window* for clearances be offered for all foreign investment applications as most other developing nations have already done.

An additional issue is whether the rising tide of applications is a part of the *'foot in the door'* strategies of foreign investors — remembering that the approvals are for *applications* only. Will these approved projects really materialize into investments in the near future? To speculate on this, we need to answer four questions that foreign investors have on their minds:

(a) Have the reforms reached the ground level, yet?

(b) Will the infrastructure facilities like power ports and rail be reliable enough?

(c) How does India compare with other options of investments in the world?

(d) Are the reforms irreversible?

The answer to (a), as discussed below, is perhaps a 'not yet'! The answer to (b) depends on the geographic location selected as well as the pace of *further reforms* in the infrastructure area. The answer to (c) is that a massive virgin market with such cheap factor resources needs more than a 'foot in the door' to be able to cash in later on. So, now is the time to start *investing,* not just applying. The answer to (d) we hope to offer later on in this paper.

In other words, the FDI angle of the 'core reforms' has received a good start. In fact the enthusiasm, according to some, may have even gone so far as to tilt the playing field in favour of foreign investors. But, whether investments will still materialize, remains an open question.

The Ground Level Reality of Regulations

Unfortunately, the environment of controls and inspections is still terribly murky at the ground level. This became fully evident in a thumb nail account put together by one of the business Chambers.[27] The staggering number of undue regulations, laws and procedures faced by industries continues to thrive, despite reforms at the top. A selected list from the above mentioned account produced below leaves no doubt on the real state of affairs at the ground level.

WHAT AN INDUSTRY STILL FACES AT THE GROUND LEVEL: ONLY A GLIMPSE

	Regulations & Complications
IDR Act & Licensing	
Clearance	Still requires clearance from 18 different government agencies.
Foreign Collaboration	Prescribed Form runs into 14 pages with 60 queries.
Industrial Laws	
Number of Statutes	51 Statutes covered in 2030 pages. Statutes require maintenance of various registers, filing of various notices, visits of labour inspectors, etc.

27 Confederation of Indian Industry (CII). An internal thumbnail account on the ground level environment for businesses for the 'Task Force On Administrative Reforms' of the CII.

Inspectors	In all the 51 Statutes, the Act empowers the inspectors to visit the factories. Almost every day, 2 to 3 inspectors are found in the factories.
Companies Act	
Payment of Fees	Event basis, 14 times.
Forms (Act and Rules)	160
Number of separate Registers to be maintained by companies	16
Prospectus	Information required to be furnished runs into 9 to 10 pages, which cover approx. 50-55 questions.
MRTP Act	
Sections	67 – even after attempts have been made to reduce the number of sections through the New Industrial Policy.
Approval under MRTP for Dominant Undertakings	50 questions including 22 sub-questions. Clearing agencies involved – 3.
Foreign Exchange Regulation Act (FERA)	
Number of Sections	81
Direct Taxation	
Number of Sections	298
Central Excise	
Number of Rules	385
Multiplicity of Tax Rules	Tariff rates given in 96 Chapters.
Power Sector	
For setting up a project in the Power Sector – clearances required	17 types of clearances, from 19 different Agencies of the government.

The above table speaks volumes on the reality, as against the rhetoric, of running an industrial unit in India, even *after* the current reforms: and it is not to be forgoten that there is another whole range of *allied regulations* and *State level* interventions that we have not even alluded to, here. Therefore, there can be no doubt that further and *even more radical changes* are called for, to *sweep out the regulatory cobwebs at the ground level,* in order that the nation can reach its new goal of a market-centered economy.

Infrastructure a Key Test for the Current Reforms

Repeated studies in the past have come to the conclusion that the failures of infrastructure translate quickly into industrial stagnation in India. In fact, the most important empirical work on the causes behind industrial stagnation during the mid-1960s–late 1970s concluded that the marked slowing down in public investment in infrastructure sectors (rail, power and coal) and the worsening management of these sectors, were among the *key factors* responsible for the stagnation.[28] Even if everything else moves in the right direction, infrastructure constraints alone can become the *Achilles' heel* for the current reforms, choking the system with supply bottlenecks.

(a) Power Sector: A Good Start, but More to be Done

There is no doubt that reforms in the power sector have forged forward more strongly than most other sectors. The Indian Electricity Act 1910 and the Electricity (Supply) Act 1948 were amended in October 1991, to encourage greater participation by privately owned enterprises at all levels of the power sector. It is noteworthy that only 4.1 per cent of India's installed power-generating capacity is in the hands of the private sector. Now, 100 per cent foreign equity is permitted in this area for an initial licensing period of 30 years extendible by 20 more years and an allowance of a 4:1 debt equity ratio. Also, there is no capping on profits for generating companies and they are exempted from the provision of balancing export earnings with repatriation of dividends. These are, by past standards, almost revolutionary changes, indeed![29]

28 See, Isher J. Ahluwalia, *Industrial Growth in India: Stagnation since the Mid-Sixties,* 1985.

29 See, *India's Electricity Sector – Widening Scope for Private Participation,* Government of India, Ministry of Power & Non Conventional Energy Sources, Department of Power, 1992. It should be added that some of the credit for the radical nature of reforms in this sector must go to the then Power Secretary, whose commitment to liberalizing this sector was extraordinary. Sometimes, individual players do tend to matter in the milieu of change.

The results from these reforms are already coming in. Private sector projects in various stages of progress amount to 8,162 MW with an investment of Rs 200 billion, and commitments from abroad already amount to $ 4.4 billion. However, what may yet create a disincentive is the number of clearances still necessary to set up a power project — 17 types of clearances from 19 different government agencies at the ground level![30] Furthermore, there is the nagging question of power distribution. The stranglehold of the State Electricity Boards (SEBs), running at grossly inefficient levels of plant load factor (approximately 55 per cent) and losing close to Rs 50 billion annually, has to be somehow broken. The politically sensitive issue of subsidy to farmers and the theft of electricity all around is linked to the SEB question, and will have to be faced sooner or later. In other words, there are still miles to go before the nation can rest in peace in this front.

(b) Telecom: Holding Back Some Key Reforms

The Government is aware that giant steps have been taken in the area of telecommunications technology in the world, leaving the old categories of 'voice transmission, switching and terminal equipment' far behind. Using fibre optic cables, communication satellites, a variety of signal-compressing techniques and digital technology, the field of data processing has been merged with the field of telecomunication in most countries. On the other hand, the age old economic theory of natural monopoly, (due to which only one company, private or public, was given charge of the national telephone systems in the West) has broken down. The new theory of 'contestable markets' recommends competition, even in the basic telephone service now.

Unfortunately, India woke up to all of this rather late and the reforms in the telecomunications sector have been moving at a snails pace. The Report of the Athreya Commission, which had recommended the conversion of the department of telecomunication's field operations into six public sector enterprises, is languishing since March 1991! Another Committee constituted to suggest amendments to the 107 year old Indian Telephone Act of 1885, is yet to submit its report. The only area that has been opened up to private sector participation hesitatingly, is the area of telecomunications

30 Ibid. p. 9

equipment, and then too, big ticket items like switches are just in the process of being tendered.

There is some serious talk of value-added services like cellular phones and paging operations being given to the private sector. But, unfortunately, *there is total antipathy on the part of the telecom bureaucracy towards bringing competition to the area of basic service.* Yet, the Government monopoly has just made a staggering *profit* of 38.55 per cent of gross revenue in 1991, using its pure monopoly power against the hapless consumer.[31] It can even be argued that such a hefty monopoly profit represents the *confiscation* of the savings of consumers without paying any dividend or interest against it.

It is noteworthy that the Minister of State for Telecom, who is known to be *pro-active and a liberalizer,* claims that the contribution of one telephone line to India's GNP is as much as $3,700.[32] If that is the case, what is holding us back from bringing in some competition and the much needed resources from the private sector into the basic service and meeting the massive demand quickly? After all, the waiting period for a telephone has grown to over seven years in the nine major cities of India and the Eighth Five Year Plan has allocated only Rs 250 billion for a projected resource need of Rs 400 billion. Expending even the latter amount would bring the current '7 phones per 1000 population' up to '16 phones per 1000' by 1995 — too small for the needs of a growing economy even by developing country standards. Obviously, we are heading towards the case of 'too little and too late' in this vital sector.[33]

(c) Petroleum: After much Struggle, some Reforms

The current situation in the petroleum and natural gas sector is not only alarming but can be expected to get worse before getting better. The output of crude is projected to *fall* dramatically once again to a low of 30 million tons and our self-reliance in crude will fall from 73 per cent in 1984-5 to well below 60 per cent.

31 See T. H. Choudhury 'Financing and Pricing of Telecom', table 3. Unpublished paper, dated July 1992.

32 See, Speech of the Minister at the National Seminar on Privatization of Telecommunication, 19 August, 1992.

33 See, Amit Mitra 'Telecom: Dynamics and Policy Shifts', *Business India*, September 14-27, 1992, for a brief essay on Telecom theory and policy.

Oil accounts for 42.6 per cent of all commercial energy used in India and often plays the crucial 'swing' role for meeting shortfalls in other forms of energy. Furthermore, the demand for petroleum products is expected to reach 77.7 million tons by 1994-5 from the current 55 million tons. The gap between what we produce and what we need will lead to severe balance of payments problems. Yet, *the government had been dragging its feet since November 1991 in liberalizing this sector fully.*[34] Only in August 1992 has it finally acted.

Small and medium *proven fields* have been opened up now to the private sector, domestic and international. Foreign investment has been allowed in three grassroots refineries, each with a capacity of three million tons: and the Fourth Round of biddings for oil exploration for 72 blocks are just completed. If the tempo of reforms continues, the long-run picture for the oil sector may be bright. But, the *short-run picture is very grim* and may have a dampening effect on industrial growth or simply balloon our balance of payments deficit.

(d) Railways: Still in the Old Tracks; and Ports: Potentially under Stress

It is unfortunate that the planners of India at one time failed to see the vital role of the railways in sustaining and promoting economic activities. They cut the Plan outlay for the railways from 23 per cent of public sector outlays in the Second Five Year Plan to as little as 5 per cent in the Sixth Plan. Add to this the gross inefficiencies, poor management, labour troubles, and what a researcher has rightly called 'the gross neglect of the maintenance of the existing capital stock on the part of the railways.' And, you have a recipe for undoing any industrial reforms!

Unfortunately, the rail bureaucracy is refusing to liberalize at a faster pace. They do not want to make their 'departmental undertakings' — the massive production units — *transparent.* They do not want to source large ticket items from the private sector competitively. They are willing to go to the extent of turning down a $300

34 See, Amit Mitra, 'Scuttling the Nation's Future', *Business India,* May 11-24, 1992 for the situation prevailing just before the reforms.
Also see the answer to an unstarred question no. 291 raised by Viren Shah in the Rajya Sabha on April 29, 1992 where the Minister of Petroleum admits that a decision in principle had been taken in November 1991 to open the small discovered oil fields to the private sector. But, 'Detailed modalities for making the offer are [still] being worked out' *(sic)*!

million loan offer from the World Bank (recently) to protect their current manner of functioning. Therefore, the reforms must now push the railways to liberalize much more deeply. Or else, not only industrial goods, but coal to thermal power stations would be held up. This dangerous interlocking mechanism between power and railway implies that if the railway fails to deliver so does power. And if the power sector fails, so does the industrial sector: thus, the vital need for immediately stepping up the liberalization of the railway system through corporatization.

The ports are a vital link to the globalization of the economy — something that the new policy purports to do. It has been rightly claimed that if the volume of trade rises by 25 per cent, nearly all of the port facilities will choke.

Therefore, major reforms in this area are called for to bring in international expertise and capital to increase the capacity of ports quickly. Unless the government acts on it now, export growth will fail to take effect in the future, despite other positive developments in the export sector.

Policy on Human Resources, Redeployment of Labour and 'Exit' Policy

From the fifties to the seventies just as *entry* into industries was controlled through licensing, *exit* from industries was practically banned through the labour laws. It was only during the 1980s that some attempts were made to weaken the controls on industrial *entry*, but *no* attempt was made to introduce *exit*. In fact, with the introduction of the Sick Industrial Companies Act (SICA) in 1986, matters got worse. Charges were flung that owners who were unable to manage their companies, found the SICA a haven for unloading their liabilities. And if the company was taken over by the Government, then, the original criteria of State ownership (discussed earlier in this paper) was violated. Unfortunately, the formation of the Board for Industrial and Financial Reconstruction (BIFR) in 1987 did not make matters any better. The crucial issue of the fate of the workers, who were faced with the closure of their company, remained unaddressed. In fact, it can be persuasively argued that not having a legal structure for 'exit' was perhaps, *more injurious to workers* than to the owners. This is substantiated by the fact that out of 100 mills that 'closed' in the city of Ahmedabad, in only 3 did the workers receive any compensation. Workers simply had no legal recourse to seek compensation, since there was no real 'exit' policy. Yet, owners have been found to precipitate *de facto* closure of factories by simply not paying the electricity bills (which is soon followed by a power cut and *de facto* closure). The owners are forced to adopt this strategy since they too have no real legal recourse to 'exit'.

Furthermore, it has been correctly pointed out by Isher Ahluwalia that 'as the Government eases the restriction on entry [through liberalization] there will be a rush for investment in new areas and mistakes will be made. The exit option assumes greater significance in this context'. In other words, if the exit policy issue keeps festering, the market-centered policies of the Government themselves may soon produce an unmanageable problem: and add to this the cases of the 56 chronically sick industries, still awaiting a solution from BIFR!

At the heart of the issue, of course, is the nature of compensation for workers in the advent of a closure of a factory, and the strategy to redeploy them. The National Renewal Fund, with a purse of Rs 20 billion, is only a necessary condition for a solution. The sufficient condition is a tripartite settlement; something that has eluded the nation so far. However, there is some positive thinking among the rank and file of workers in favour of the solution of a 'golden handshake', provided the size of the 'handshake' is reasonably large. And this is reportedly true of even the more militant members of the Communist trade union, CITU.

But, experience in Gujarat shows that a package which includes some *income-generating options* from the 'golden handshake', is the most effective solution, particularly, if retraining is *also* a part of this package. Of course, the ultimate problem lies in producing a national consensus between employers, trade unions and the Government, where the trade union leaders do not find their own self-interest (of leadership and patronage) at stake and the Government is not perceived as having let down the interests of the workers. With the lapse of time, there has been a softening of attitudes, but a way to clinch this problem is yet to be found.

Reorganization of Government: Role, Structure and Functioning

The politico-economic underpinning of the new reforms obviously demands a total reorientation of the role, structure and functions of the Government. This is a very challenging and thorny task, given the crystallization of the State-centered system over fifty years.

The *role* of the Government will have to shift from that of a super controller, the super restrainer and *the* director of economic activity, to that of a *facilitator* and *nurturer* of private productive forces. The most important tasks in this new role are:

(a) remaking rules that are totally *transparent with no discretionary powers to individuals* : rules that are simple and offer a *level playing field;*

(b) enforcing such rules effectively;

(c) enforcing contracts and clearly defining and upholding private property rights;

(d) allocating resources for the construction of *infrastructure*(quasi-public goods) and allocating resources for building up *'human capital', without owning and operating* such facilities in principle.

There is certainly some movement in this overall direction under the current reforms. The tyranny of the *case-by-case* basis has been reduced and transparency has been kept clearly in mind. Small signals, like the recasting of the tenancy laws in major cities, are offering hope for the enforcement of property rights. Changes in the infrastructure area, already discussed, also point to some movement in the above direction, though slowly, perhaps. The area that has *not* been addressed so far is the crucial area of human capital formation at a much faster pace. This gap has to be filled quickly. Also, the modalities of enforcement within the new structure of rules lies unaddressed and somewhat chaotic. As a senior bureaucrat put it in a private conversation: 'We still do not know how we really ought to respond to violations by private parties in this new milieu of liberalization'.

The current structure of Government will have to be severely pruned and the mechanisms to do so are too complex to discuss here. However, the *functioning* of the Government machinery in this fast changing environment needs our attention. It has been rightly pointed out by a thoughtful insider, to the Government that, 'Government intervention in industry has to be far more knowledge-based'[39]. Rakesh Mohan concludes poignantly that for optimally intervening in certain complicated areas like tariff and excise structures, 'at present, the Government does not possess the knowledge-based machinery required to accomplish such reforms effectively'. The recent securities scam, which came after Mohan's reflections, has proved his point: therefore, Mohan's reflections need immediate attention.

Reorganizing Business: From Mercantilism to Competition

It is clear that larger businesses in India developed in an environment where the key to one's success lay in the cornering of licenses and working through the maze of Government regulations. We have already demonstrated that entry, expansion and even day-to-day functioning was centred around licenses and regulations. A significant amount of time, managerial skills and resources went not into increasing efficiency and production in industry but into 'wheeling and dealing' with the 'license-quota-control' *raj* successfully. Preventing the entry of competitors was not through higher efficiency and innovation but through contact networks within the Government which provided timely information on licensing plans and ways to secure the licenses. Monopolies or oligopolies could be set up through 'rent giving' and oiling of networks within Government and meeting the needs of electoral funds of politicians. Furthermore, the policy of punitive import substitution provided a perfect shelter for Indian industries from any kind of foreign competition, making redundant the acquiring of the state of the art technology or knowledge. This behaviour pattern of the industry was perfectly natural, given the institutional structure within which they were called upon to function.

Now, in this *new* milieu, with much lower levels of entry barriers and regulations, and greater chance of foreign competition, the primary thrust of business management *will have to shift radically towards efficiency and innovation.* Also, with the inevitable invasion of new technologies (due to the liberalized policies on technology transfer), businesses would be forced to make more complicated choices (based on specialized knowledge) at all levels of production. In fact, the current trend of fragmentation of the family businesses of giant Indian industrial houses may well have its roots in this emerging environment, as well as in the mindset of a new generation within these families who were trained professionally in the West from an early age. Their vision of the world and excellence in business is quite different from those of their previous generation.

Fragmentation of these large business houses has, of course, a negative side. Their 'economy of scale' advantage could be reduced, making them more vulnerable to competition from much bigger multinational firms. On the other hand, since the structure of their businesses was partly *shaped by the regulatory framework of the Government,* the new liberalizing environment is bound to take some natural casualties.

Perhaps, the positive outcome of this hiatus may well be a new structure of collaborative businesses with international firms, based on the state of the art technologies and managerial functions. If that were to become possible, the fragmentation of large business houses would turn out to be a major blessing in disguise. However, *the playing field for Indian and foreign businesses would have to be even* – something that is supposedly tilting, according to critics, in favour of foreign firms in some areas currently.

An important issue that needs deeper reflection today, is the integration of small, medium and large business firms, as is the case in Japan and East Asia. Since the Government has not yet begun thinking on this issue (shown earlier), businesses themselves must take the lead in thinking through this question. It is quite evident that there is a large pool of entrepreneurial energy in the medium, small and informal sectors. Yet, these sectors are constrained by a variety of factors, including market information networks, quality control systems and most importantly, lack of property rights and legitimacy, particularly, in the informal sector. With an easing of these constraints, the profitable links with larger businesses might escalate, benefiting all of the concerned sectors.

Also important is the need to forge deeper links with the already modernizing agricultural economy. Here, instead of the Government leading the Integrated Rural Development Programmes (IRDP), it is the burgeoning corporate sector that must take the lead. Experiments like the Etah project of Hindustan Lever, the Rural Support Programme of Lupin, Mafatlal Group's IRD programme and Tata Steel's Rural Development Society need to be multiplied manifoldly.

Finally, the forging of deeper links with the workforce is very crucial in the 'new' India. As the neo-Marxian perspective of an *antagonistic class relations fade* into the background, new perspectives on worker-owner relationships must replace it. In this crucial search and evolution, based on traditional as well as modern structures in India, the captains of industry cannot afford to take the back seat. In other words, the time has come to not only restructure businesses internally in order to meet the new policy environment in the country but also to write another 'Bombay Plan' of the 1990s to meet the new challenge.

REFORMS WHICH COULD BOLSTER OR RETARD INDUSTRIAL GROWTH: FINANCIAL, FISCAL AND TRADE REFORMS

In order to reach a higher trajectory of growth, industry needs investible funds; it needs a tax structure that allows a reasonable rate of retained earnings; and it needs a trade policy which promotes export in order to import raw materials, capital goods and technological expertise. Severe bottlenecks in any of these three crucial areas could retard industrial growth and withhold industrial transformation. Therefore, we wish to touch upon financial, fiscal and trade policy reforms only to the extent that they affect the industrial sector. Needless to say that full papers on each of these concerns are necessary and FICCI is organizing conferences on each of these topics where such papers are expected to be presented. So, we restrict our comments below to merely raising some basic questions and seeding some preliminary ideas.

Financial Reforms: The Urgency for Implementing the Narasimham Committee Report

Financial institutions and policy instruments in modern nations are expected to (a) facilitate the mobilization of investible funds and (b) promote the efficient allocation of these funds. India, fortuitously has a savings rate of 22 per cent of GDP today and the private households sector provides over 80 per cent of these savings. This belies the thesis of the 1950s-1960s that poor nations cannot save and therefore must rely on foreign aid to push themselves out of the 'low level equilibrium trap'.

There is no doubt that India has succeeded over these decades to not only save a large proportion of GDP but also to channelize much of these savings through the formal market structure via a variety of financial intermediaries. Furthermore, a significant number of rural households now interact with modern financial intermediaries as well. However, the key problem with our financial policies has been the highly inefficient allocation of the investible funds — policies which have crippled our industrial growth and development.

The misallocation of the generated funds fall under three categories:

(i) Policy induced ('forced') allocation of investible funds into public sector enterprises (PSUs).

(ii) Policy induced 'drain' of large amounts of investible funds for meeting massive budget deficits.

(iii) 'Forced' or 'directed' lending of investible funds to the so called 'priority sectors' at highly subsidized rates of interest.

The first two types of inefficient allocations take place through the financial instruments devised jointly by the Reserve Bank of India and the Banking Division of the Finance Ministry. These instruments are the 'statutory liquidity requirement' (SLR) and the 'current cash reserve' (CCR) requirement, both of which have been kept at inordinately high levels. For example, a bank is compelled to place as much as 53 per cent of its entire deposits with the Reserve Bank of India or in government papers in order to meet the SLR and CCR requirements. Furthermore, a bank is compelled to lend another 20 per cent of its deposits to the so called 'priority sectors' of the economy. Obviously, a very small proportion of bank deposits are left for private commercial lending. In other words, *private commercial borrowers are 'crowded out' of India's bank lending market due to the financial policy parameters of the Government.* Strongly criticizing these policies, the Narasimham Committee Report states that the SLR and CCR have, in fact, become '... major instruments of mobilization of the household sector's financial savings in the form of deposits with the banking system to finance the public sector' [*sic*].[35]

Furthermore, the interest received from the SLR and CCR barely break even against the costs incurred by banks towards such lending. Of course, the interests from lending to the 'priority sectors' is not even intended to break even, since this sector was meant to borrow at highly subsidized rates of interest. Therefore, the total burden of

35 See The *Financial System Report* by M. Narasimham (Nabhi Publicaion, 1992) p. 38.

cross subsidy falls entirely on commercial borrowers — a group that is small and rapidly shrinking. No wonder that a recent study by the World Bank states that 'commercial borrowers have been penalized by having to pay high real interest rates to sustain government borrowing and priority sector credit.[36] Given the low rates of returns from SLR/CCR and the burden of subsidies, it could be argued that the recent involvement of banks in the 'stock market scam' (to make a quick buck) may have been precipitated by the internal need of banks to make profits from somewhere in order to compensate for its drain into SLR/CCR and priority sector lending.

After studying these issues carefully, the Narasimham Committee Report recommends the gradual winding down of SLR, commensurate with the cutting back of the budget deficit. This should 'enable banks to allocate their resources on a flexible and efficient basis to promote investment and production in agriculture, industry and trade with due regard to the productive use of their resources', says the Committee.

On the issue of 'directed credit programmes' (to the 'priority sectors' of the economy), the Narasimham Committee is of the view that '...the pursuit of distributive justice should use the instrumentality of the fiscal rather than the credit system'. In fact, the Committee finds that the compulsory 'directed credit programmes' have '...led to segmentation of credit markets and introduced an element of inflexibility in bank operation' [*sic*].[37] The Committee goes further in concluding that the whole package of financial policy instruments imposed on banks (as discussed above) '... has meant that the traditional sectors of medium and large industry and trade have often been subject to severe restriction with consequent deleterious effects on the growth of industrial investment and production.[38]

What could be more forthright and clear in demanding a radical change in policies? Yet, it is unfortunate that most of Narasimham Committee's recommendations have remained unimplemented to this day. The purpose of the Report was to 'ensure that the financial services industry operates on the basis of operational flexibility and

36 See the recommendations of this study in *Economic Times*, July 5, 1991.

37 Narasimham Committee Report, op.cit., p. 39

38 Ibid, p. 40

functional autonomy with a view to enhancing efficiency, productivity and profitability'.[39] If these goals are not met soon, no amount of industrial policy reforms will succeed in producing the desired result of economic resurgence.

Fiscal Reforms: Awaiting the Full Implementation of the Chelliah Committee Report

Fiscal policies of nations attempt to garner resources from the economy to pay for different functions of governments. However, tax rates fixed at exceedingly high levels often result in lower tax revenues — the proverbial Laffer Curve effect. High tax rates also deter the production of goods and services and drive earned incomes underground into the illegitimate ('black') market. India's average tax rate is known to be inoptimally high and has produced most of the above mentioned adverse effects. In fact, lowered income tax rates in the mid-1980s produced an increase in tax revenues, confirming the Laffer effect for the case of India. Furthermore, the base of taxation in India continues to remain extremely narrow. With these problems in view, a 'long term tax policy' (LTFP) was announced in 1985 by the then Finance Minister, Mr V. P. Singh. But, with Singh's departure from the government, the LTFP was also discarded.

With the new wave of economic reforms in late 1991 a Committee on Fiscal Reforms was appointed by the Government under the chairmanship of Raja Chelliah. However, some recommendations from the First Report of the Chelliah Committee found their way into the Budget of 1992-3 (in March 1992). The Government has accepted the main thrust of the Report in principle, but major recommendations still remain to be implemented. Chelliah expressed concern over the fate of his Report by stating that the 'implementation of all the changes mentioned in the Report may be difficult because of political considerations'.

Aside from lowering the rate of taxation, raising the income tax exemption ceilings and simplifying the procedures, the Chelliah Committee also attempts to redirect the flow of savings — something of importance to the industry. The previous areas of tax concessions were primarily those from which the Government

39 Ibid, p. 2.

could easily draw resources and invest in public sector undertakings or use the resources to meet its budget deficit. By recommending the abolition of many of these concessions, the Chelliah Committee, attempts to direct savings into the capital markets of India. There is no doubt that the competitive nature of India's capital markets will compel it to use resources more efficiently than the Government, and India's industries would have an additional source for raising investible funds.

The Chelliah Committee makes a crucial recommendation on wealth tax that could provide a major incentive to India's industries. 'The tax on wealth should be abolished in respect of all items of wealth other than those which can be regarded as unproductive forms of wealth... 'says the Report. However, given the socialistic mindset in the country, the Government has hesitated so far in implementing this 'politically sensitive' recommendation. Similarly, the Chelliah Committee has recommended a reduction of the general level of tariff, reduction of the spread or dispersion of tariff rates, simplification of the tariff system, and abolition of the 'practice of making changes in effective rates through notifications'. If these recommendations on tariffs are implemented fully, industry will see greater competition and tend to become more efficient over time. Such changes could also increase the capacity of Indian industries to compete in the world market of exports. The rate of reduction over the time specified by the Chelliah Committee is crucial for the success of these fiscal reforms.

For domestic production of services and commodities, the Chelliah Committee is in favour of relying on value added tax (VAT) 'with two or three simple rates say, at 10, 15 or 20 per cent'. Selective excise duty 'on non-essential consumption could be levied at 30, 40, or 50 per cent', not exceeding 50 per cent. These recommendations represent a wide consensus in the country and need early implementation. There is no doubt that fiscal reforms are crucial for liberalizing India's economy, and equally so are stimulating industrial growth and structural transformation.

Foreign Trade Reforms

Foreign trade policy in India through the fifties to the seventies was firmly rooted on the principle of 'import substitution' and suffered deeply from export pessimism. India stuck to a highly 'restricted trade regime' through these decades, using high tariff walls and a variety of quantitative restrictions. Interestingly, the only study which had argued persuasively against export pessimism in the 1960s was authored by Manmohan Singh, now the Finance Minister of India, liberalizing India's economy![40] Unfortunately, Singh's pleas fell on deaf ears till the end of the 1970s.

The 1980s saw some rethinking on trade policy in India with a degree of liberalization of capital goods' imports during the second half of the 1980s. This did lead to some technological up-grading in India's industries. However, tariff walls still remained exceedingly high. For example, the nominal tariff in the manufacturing sector in 1987-8 was as high as 146 per cent, when the average for most developing countries was as low as 25-30 per cent The average duty collection rate on total imports, after taking account of exemptions, was 60 per cent for India in 1987-8.[41] But, despite the negative trade policies of the Government through the 1960s and 1970s and partly due to the liberalizing attitudes of the 1980s, the share of manufactured goods in total exports rose from 45 per cent in 1960-1 to 75 per cent in 1989-90. In other words, the potential for modernizing our exports was very much there all through.

The first major impact on the policy parameters on trade in the 1990s came with the devaluation of the rupee by 23 per cent in July 1991. This was followed by several liberalizing drives on the part of the Minister of Commerce, P. Chidambaram. The key policy that changed the operations in the foreign trade sector and continues in force, is the so-called 'partial convertibility of the rupee' in the current account. On 1 March 1992, the Finance Minister announced in the Budget that 'all foreign exchange remittance, whether earned through exports of goods and services, or remittances, will be converted into rupees in the following manner: 40 per cent will be converted at all official exchange rate,. while the remaining 60 per cent will be converted at the market determined rate'. In other words, the exporter will stand to receive a premium (officially) from exports which now stands at rupees five per dollar of exports. In

40 See, Manmohan Singh, (1964), *India's Export Trends.*

41 See, Isher J. Ahluwalia's *Exim Bank Commencement Day Annual Lecture,* 1992, for a brief but clear discussion on trade policy in India.

addition, the new rules permit an exporter to retain 15 per cent of export earnings in hard currency in an Indian bank.

Furthermore, the new policy also stated that while the foreign exchange surrendered at the official rate will go towards the import of essential commodities, 'all other imports of raw materials, components and also capital goods will be made freely importable on Open General License, but the foreign exchange for these imports will have to be obtained from the market'. This new policy eliminates the decades old licensing structure for imports, leaving the volume of imports to be regulated by the available volume of foreign exchange. Of course, the price of foreign exchange will now provide the market signals for equilibrating the supply and demand of foreign exchange. Only, a small list of items have been held back in a 'negative list' which will require an import license.

While quantitative restrictions have been, by and large, removed, the peak level tariff has also been reduced to some degree. It has been lowered to 110 per cent except for passenger baggage and alcoholic beverages. For capital goods imports, a new scheme entitled the Export Promotion Capital Goods (EPCG) Scheme has been introduced to facilitate imports at a concessional duty of 15 or 25 per cent of CIF value. This scheme covers second hand capital goods and components as well. In addition, special emphasis has been given towards liberalizing the rules and regulations for the 100 per cent Export Units, Free Trade and Export Processing Zones.

There is no doubt that the Government has taken some bold steps in the area of trade reforms. If not anything else, India has finally shed its age-old inward looking 'import substitution' approach to foreign trade. Punitive physical controls over imports have given way to fiscal controls. Price signals for allocating foreign currency have come to be respected. A major part of the bureaucratic labyrinth has been cut back. But total convertibility, at least in the trade account, still remains illusory. And the trade reforms have lost the key liberalizer, Mr Chidambaram, who resigned his post as Commerce Minister in the wake of the 'securities market scam'. It must be admitted that at certain points of history, quality of leadership does play a crucial role. The unfinished tasks left behind by Chidambaram in this area are many and new leadership is still not in place. It can only be hoped that the momentum of reforms will push towards completing these tasks and further liberalizing India's trade sector.

SUSTAINABILITY OF CURRENT REFORMS

The key questions that observers the world over have been asking are: Are these reforms sustainable? and, can these reforms be reversed? From the range of issues discussed in this paper, (though pertaining to the new industrial policy only), some speculation on these two vital questions can be offered.

First of all, there is no doubt that unless the labyrinthine control mechanism at the ground level (discussed at length in this paper) can be swept aside quickly, these reforms could flounder. The State and local Governments will have to be roped in and a much larger campaign against these ground level control mechanisms will have to be mounted. For this the Government alone is insufficient. A variety of forces at the grassroots levels will have to be combined, along with businesses of all sizes. The task is truly daunting indeed!

Secondly, these reforms have come entirely from the top. In fact, there are very few politicians, even in the ruling party, who really understand let alone fully subscribe to the underpinning of these reforms. Furthermore, the ethos in the country, nurtured over 50 years or more, is hostile to a market-centered growth process. The inherent suspicion of our people about business activity as being another form of theft — a suspicion that permeates the mindset of our intelligentsia — makes these reforms sit uneasily on our soil. So, it is not surprising that the word 'capitalism' *continues* to be a 'dirty word' in the political lexicon. Also, foreign investment is still viewed with suspicion. The ghost of colonialism has not been exorcised yet.

After all, *ideas and perceptions* rule the world to some extent. With such a difficult ethos base, even a short run economic crisis, like a sudden escalation in consumer prices, can create significant difficulties for these reforms. The *dirigiste* forces can quickly capitalize on them. It should be noted that the recent securities scam did not have this kind of an electric negative effect because of the narrow base of India's capital market in relation to the population size.

Unfortunately, industrial growth was negative in 1992 and still refuses to pick up substantially. There were six reasons offered for this by the Government in February, 1992.[42] Most of these were

42 See, *Economic Survey*, 1991-2, pp. 83-4.

policy-induced reasons and have been reversed since then. Now, if industrial growth does not pick up, pressures will mount on the Government to slow down the reforms.

Similarly, exports had suffered during the process of belt tightening and from changing the foreign exchange policies. They too must pick up. Fortunately, there are some hopeful signs on this front. In other words, we are getting to that point where these vital economic signs must start reflecting the positive impact of the reforms. Failing this, tensions could mount and difficulties could arise in sustaining these reforms.

CONCLUDING REMARKS: THE TASK AHEAD

In this paper we have traced the evolution of the industrial policies in India through history, keeping an eye on the political economy behind these policies. In so doing it has become clear that the roots of the State-centered model followed since Independence are extremely deep and cannot be extricated easily or recast readily. The 'battle' to change will have to take place at many levels, involving the transformation of the very way a nation thinks. Therefore, the road ahead is hard and murky. The most difficult of the tasks ahead is the revamping of the ethos that has supported the edifice of the past policies for over fifty years. Then comes the nuts and bolts problem of sweeping out the plethora of controls and regulations at the *ground level.* What has happened so far are mere microlevel changes in 'core policies' of the Central Government. The frightening description of the ground level realities provided in this paper warns us against complacency. It warns us against underestimating the depth and spread of the 'license-quota-permit-inspector' *raj.* The hope, however, lies in the fact that both economic theory and recent international history is on the side of those who seek change. Time, of course, will be the ultimate judge.

POSTSCRIPT

Since the completion of this paper, India's policy-makers have pushed the nation further and deeper into the process of 'structural adjustment' (liberalization of its economy). The Annual Budget of 1993-4 exuded self-confidence and announced 100 per cent convertibility of export earnings into rupees on the basis of the *market defined* rate of exchange. The Government (the Reserve Bank in this case) will not have to fix an arbitrary rate of exchange any more. This also means that the Government will not have at its command 40 per cent of the hard currency earnings of the nation as it did in the 40%=60% formula followed since the 1992-3 Budget. The practical implication of this is that organizations like the Oil and Natural Gas Commission (ONGC) will now have to purchase oil from international sellers at a market driven exchange rate, (perhaps 20 per cent higher than the price paid earlier), forcing public sector (or the so-called autonomous) organizations to tighten their belts. They may even have to soon restructure themselves to cope with this squeeze.

Of course, the exporters stand to gain significantly from this market-based exchange rate policy. They would be able to collect a larger premium from their export efforts. However, the nation is expectantly waiting for *real* convertibility — a regime in which Indians will have the right to hold their export earnings in hard currency or for that matter, be able to convert their rupee earnings into any currency at any time. At such a stage, the policies of liberalization will have reached even a higher plane of maturity. Unfortunately, Indian economists who favour this step are still divided on the timing of moving to real convertibility. Such a division among liberalizers has led to some hesitancy on the part of the Government to move to this ultimate step.

On the licensing front there has been further progress too. Some of the eighteen industries kept under the paw of the licensing *raj* have finally found freedom. Automobile, electronic goods, leather and paper industries have been taken out of the licensing list. But the remaining few still stick out like a sore thumb. Can fiscal means not be found to control them, if control them at all? Why not simply cut out the cancer of the 'licensing *raj*' by abolishing the system of licenses altogether?

In the crucial area of infrastructure, potentially the *Achilles' heel* for India's reforms, there has been further progress. In the power generation sector, foreign direct investment did not come in droves despite the sanction of 100 per cent foreign equity. The obvious reason was the concern of foreign investors of non-payment by the State Electricity Boards (SEBs). These SEBs are notorious for their massive losses in power transmission and have turned a blind eye on the blatant thefts of power by consumers, often in league with SEB officials. This fear on the part of investors has now been allayed through a sovereign guarantee by the Central Government of a 16 per cent return on capital invested in power generation.

Furthermore, a predatory demand by the railways on power sector investors that they must own their own wagons (for moving coal) — in effect, a demand for an upfront payment of $ 100 million or so for a medium sized power generation unit — has been successfully quashed by the Government. Similar predatory demands by the public sector monopoly in coal has also been quashed. What still remains are the 17 regulations, 13 of which are statutory. Even worse, these permissions will have to be sought from 19 different departments of the local, State and Central Governments. Hopefully, a single window will soon be opened and then India will see the inflow of foreign direct investment in the power generation sector, addressing the most important area of industrial infrastructure. My recent discussions with Western law firms which specialize in the regulatory and legal aspects of international investment in power confirm the above hypothesis.

In the area of telecommunications, vital for the follow-through of India's reforms, there has been some progress. However, there has also been a lot of foot dragging, bureaucratic machinations, and even the occasional throwing of spanners on the wheels of change. The good news is that the crucial area of switching equipment has been privatized. Four international giants — Fujitsu, Siemens, Ericsson and AT&T — are going into local production of large-sized switching equipment. This will allow India to offer many more telephone exchanges and therefore, many more telephones for the consumers. It is disturbing to know that the average waiting time for the installation of a telephone in India today is seven years and the average success rate of a telephone call is 0.27 (one call in four). With these ground level realities it is high time that India had thousands of new switches and with them the latest technology in the field of switching. Hopefully, the international giants can now provide us with both.

The bad news in the telecommunication field is that the privatization effort of value added services — cellular phones and paging services — have been marred by a court case which ended with the castigation of the department of telecommunications in these unequivocal words: '... the petitioner (against DoT) has not been given fair and equitable treatment.... The action of the department would appear to be rather arbitrary and irrational, and even whimsical' [sic]. (The High Court of Delhi, C.W.P. No. 4030 of 1992; date of decision: February 26, 1993; page 90.)

Since these negative pronouncements of the High Court the Government has remained petrified, unable to solve the impasse. Such a delay may well result in the exit of some of the multinationals who had formed joint ventures with Indian firms to bring value added services in telecommunication to India. Obviously, some transparent and concerted policy action is the need of the hour in this vital sector.

There are two nagging deficiencies in the reform process which need special mention. The first is the attitude and functions at the lower levels of the bureaucracy. Going through the evaluations of the high power Foreign Investment Promotion Board may be easy and quick, but once the files reach the lower levels of Government the speed drops dramatically and 'palm greasing' begins. In other words, the contractual work culture still remains a rarity.

The second deficiency is the continued lack of an information and knowledge-based structure of governance. While the private sector in India has rapidly hitched on to the burgeoning world of informatics, even without a telecommunication revolution, the Government of India continues to rely on archaic methods and conservative mindsets. While there are a few exceptions in the North and South Blocks of Delhi, insiders like Rakesh Mohan are right in demanding a totally new ethos and style of governance. In the new milieu where the State is expected to play the role of a facilitator of private business activities rather than that of a judge and jury, information processing and a high degree of analytical skills will be demanded of the government functionaries. They, unfortunately, do not possess these skills nor has the Government yet come up with a coherent retraining plan to achieve this transformation. Of course, the hardware for such a restructuring is nowhere in sight. Therefore, India's leaders will have to develop a clear strategy to bring about this difficult and painful transformation, failing which the process of liberalization could open a Pandora's Box. Yet, there is absolutely no doubt that India has turned the corner and during the past two years of reforms the fundamental premises have been laid. Leftism is *passe* if not out all together and entrepreneurship is in. Import-substitution, the corollary of what was haughtily called 'self-reliance', is also out, and interfacing with the global economic process is in. Turning back the clock to the State-centered oligarchy ('Indian socialism') will no more be easy for any political party in the future.

Agrarian Change in India since Independence

GURCHARAN SINGH BHALLA

Indian agriculture, which was characterized by severe institutional rigidities, technological backwardness and low levels of productivity and income during the colonial period, has undergone significant changes since Independence. These changes have been brought about both as a result of implementation of land reforms, and undertaking of large investments in irrigation, power, and other rural infrastructure. As a result, appreciable acceleration has taken place in the growth rate of agricultural output and productivity in the post-Independence period, particularly after the advent of Borlaug seed-fertilizer technology during the mid-sixties. However, because of large differences in initial resource endowments and variations in the level of infrastructural investments, gains of technological change have not been equitably shared by various regions. Furthermore, because of failure of land reforms in bringing about equitable distribution of landholdings through ceilings, all sections of peasantry and landless agricultural workers have not been equally benefited.

The purpose of this paper is to briefly review the pattern of agricultural development in the post-Independence period and to discuss the changes that have taken place in the agrarian structure of India since Independence. The paper is divided into four parts. The first part is devoted to an examination of changes that have taken place in the agrarian structure of India in the post-Independence period as a consequence of land reform legislation. The second part brings out the existence of large variations in the nature and extent of the spatial pattern of agricultural development in India. The third part is devoted to a brief discussion of the emergence of new contradictions and increasing rural tensions in present day India. Finally, the main conclusions are summed up in the fourth part.

LAND REFORMS AND CHANGES IN AGRARIAN STRUCTURE: POST-INDEPENDENCE PERIOD

India was a semi-feudal and backward agrarian economy on the eve of Independence. With nearly two-thirds of the workforce engaged in agriculture and more than half of total domestic products originating in it, agriculture constituted by far the most dominant sector of the Indian economy. The agrarian structure was characterized by several institutional features that perpetuated inequality and social injustice. There existed three types of land settlements in different parts of India — the zamindari system, the *ryotwari* system and the *mahalwari* system. Zamindari settlement obtained in the erstwhile Bengal Presidency. In this area, absentee landlords were given the right to collect land revenue and they increasingly acquired title to land; intermediary tenures were most prevalent; leasing and subleasing of land were commonly practised and; the tenants had hardly any defined rights. The *ryotwari* and *mahalwari* settlements prevailed in the Madras and Bombay Presidencies and the North Western Provinces. In these areas, the cultivators or the cultivating sections of the village community were given direct land rights and, in theory, intermediary tenure did not obtain. But even there, leasing-in and leasing-out of land took place on a large scale as land passed into the hands of non-cultivators due to indebtedness and other reasons. In the Princely States of India, by and large, landlordism existed in its naked form with all sorts of lease arrangements, and tenants had no defined rights. In all parts of rural India, multifarious types of vested interests in land provided powerful support to the British Raj, but the existence of semi-feudal relations acted as the most important barrier to the development of agriculture.

The production technology in agriculture was traditional and only a limited attempt had been made to upgrade it during the British period. Some technological developments did occur as a result of integration of the Indian with the British economy, but these were primarily confined to commercial crops. With the setting up of the Imperial Council of Agricultural Research at Pusa, Bihar, in 1905, a small beginning was made in creating an infrastructure in scientific research and extension. In addition, as a belated response to recurrent famines, the British undertook large investments in canal irrigation in the late nineteenth and early twentieth centuries with a view to increasing the output of food crops. But these measures had only a limited impact on certain crops in a few regions and failed to lift Indian agriculture out of stagnation. Thus the total crop output index registered an increase of only 12 per cent from 1904-5 to 1944-5. While output in foodgrains did not show any increase whatsoever, the output index of non-foodgrains rose by 54 per cent. However, most of the enhanced output was due to increase in area. Per hectare, the productivity of non-foodgrains crops rose marginally by 9 per cent. But the total crop per hectare registered a decline of 5 points, while foodgrains output per hectare fell by a full 12 points. As a result, the per capita food availability declined from 200.2 kg per year in 1905-6 to 152.2 kg per year in 1945-6, even though the population growth rate during the period 1901- 41 was only 0.83 per cent per annum![1]

To sum up, several factors were responsible for the slow growth of Indian agriculture during the British period. Among these were the deliberate attempt to integrate the Indian economy with the colonial, and the lack of any policy for the development of resources of the national economy for the welfare of the people. Foreign competition and the restrictive policies of the Government resulted in the destruction of indigenous cottage industries and increased the pressure on land. Large-scale semi-feudal land relations posed paramount barriers to any attempt at modernization of agriculture. Added to this was the lack of adequate infrastructural investment in irrigation, power, and agricultural research and extension in large parts of the country. Even in 1944-5, irrigated area accounted for only 23.5 per cent of total cultivated area. Furthermore, about 90 per cent of the total area irrigated up to the forties was concentrated in the four provinces of Punjab, Sind, the United Provinces, and

1 India, Ministry of Agriculture and Irrigation, *Report of the National Commission on Agriculture*, Delhi: Controller for Publications, 1976.

Madras, and the rest of the country continued to depend on the vagaries of the monsoon.[2]

The post-Independence period in India is characterized by unprecedented changes in its agrarian structure. The nature and significance of these changes can only be understood in the light of the character of the Indian national movement that was dominated by the Indian National Congress. During the freedom struggle, the national movement was able to build a broad alliance of class forces ranging from all sections of the peasantry, the shopkeepers, the middle classes and the bourgeoisie. Apart from colonial bureaucracy, the main opposition to the national movement came from the big landlords, moneylenders, princes and other vestiges of the feudal order. Another complicated factor was communal fission which defied any class characterization. The national movement largely succeeded in making the anti-landlord and anti-feudal struggle a part of the struggle for national independence. In the process, it was committed to bring about radical land reforms after coming into power.

The National Government set out to implement a comprehensive land reform policy soon after assuming power in 1947. The main objectives of land reform legislation passed by the various State Governments during the mid-fifties were the abolition of intermediary interests, regularization of land rent, and provision of security of tenure and eventual ownership rights. The declared policy was also to consolidate landholdings in order to increase productivity. The achievement in the abolition of intermediary tenure was quite impressive. By 1964, all States had passed legislation abolishing landlordism. As a result nearly 20 million tenants were brought in direct contact with the State and about Rs 6700 million worth of compensation was paid to the ex-intermediaries.

However, in many parts of India, the landlords, with the connivance of the local bureaucracy, were able to resume land for self-cultivation by ejecting a large number of tenants. In many cases, delay in implementation of the Act, the lack of proper land records with respect to cultivation by crop-sharers and tenants, and conservative and often pro-landlord attitudes of the judiciary resulted in the denial

2 Hirashima, Shigemochi, *The Structure of Disparity in Developing Agriculture: A Case Study of the Pakistan Panjab*, Tokyo: Institute of Developing Economies, 1978.

of rights to tenants, thereby sabotaging the objectives of the legislation. Furthermore, in some States, particularly Bihar, Orissa, Rajasthan and Madhya Pradesh, where the peasant movement was weak, the landlords, because of their power and influence, managed to evict tenants and keep large holdings for self-cultivation. With regard to the reform of tenancy, several enactments were passed by the various State legislatures fixing rents not exceeding one-third to one-fifth of the gross produce. In this respect, the actual achievement varied from one State to another. In general, market conditions exercised greater influence than the diktat of law. Further, in most States land records were not up-to-date, and even where these were regularly rewritten, the ground level reality was not reflected in them, particularly regarding tenancy and share-cropping. This enabled landlords to perpetuate tenancy and the tenants were unable to fight for their rights.

Despite several bouts of legislation during the fifties, late sixties and early seventies, all attempts to impose ceilings on land-holdings and distribution of surplus land among the landless and poor peasants have yielded very poor results. According to the latest estimates, by 1988, 73.38 lakh acres (2.968 m. hectares) had been declared surplus as a result of implementation of land ceiling laws. However, only 60.4 per cent of the area declared surplus has actually been distributed,and about 18.3 per cent is involved in litigation. It is notable that the officially estimated surplus is only a fraction of the area held in large ownership holdings, as estimated from survey data. According to some estimates derived from the National Sample Survey, Twenty-Sixth Round (1972), the surplus area should be as high as 214.89 lakh acres (8.70 m. hectares), that is nearly three times that estimated by official agencies.[3] It is obvious that rich landowners have been able to sabotage the implementation of the ceilings' laws through the active connivance of the land revenue administration. Another significant feature is that the landless labourers who have been toiling on land for centuries, have hardly benefited from these land reforms except for the hope of getting a homestead in some States.[4]

3 India, Planning Commission, *Proceedings and Papers of the Seminar on Land Reforms — A Retrospect and Prospect*, Delhi, GOI Press; see specially Asha Swarup's Background Note, pp. 49-89, 1989

4 Herring, Ronald J, *Land to the Tiller — The Political Economy of Agrarian Reforms*

Regarding consolidation of holdings, most States have not shown any enthusiasm for it. It is only the north-western States that have been able to undertake consolidation in a successful manner.

One can therefore conclude that in terms of agrarian structure, a major change has been brought about through the abolition of intermediaries in many States. However, the land has not been given to the actual cultivators in all cases. In many cases, large-scale ejectment of tenants has taken place and big cultivators have been able to get away with large holdings. This is particularly true in Bihar, Orissa, Rajasthan and Madhya Pradesh where, from all available evidence, very little dent has been made in the age-old semi-feudal relations. Even in other States, land reform legislation has failed to reduce the wide inequality in land holdings. The National Sample Survey data indicate that the inequalities in terms of land distribution continue to be very great. Since the working population in agriculture is growing rapidly, the per worker availability of cultivable land has registered a sharp decrease and the extent of marginal and small holdings has risen sharply.

AGRICULTURAL GROWTH IN THE POST-INDEPENDENCE PERIOD

The post-Independence period marks a turning point in the history of the development of Indian agriculture. The policy-makers adopted a two-fold strategy for regeneration of agriculture. The first element of this policy consisted of changing the rural institutional structure through land reforms, and the second was the creation of a big infrastructure in irrigation and power. The growth performance of agriculture since Independence has been quite impressive. Thus, compared with a paltry rate of growth of 0.28 per cent per annum during the first half of this century, agricultural output rose at a rate of 2.69 per cent annually during the post-Independence period 1949-50 to 1990-1 compared with a population growth rate of nearly 2.15 per cent per annum. Although this falls short of the needs of the economy, it does represent a tremendous advance over the earlier historical record.

---

in South Asia, Bombay, Oxford Unversity Press, 1983. Herring gives a fascinating and critical account of land reforms in South Asia.

It is analytically useful to divide the post-Independence period into two sub-periods. The first of these roughly extended from 1949-50 to 1964-5. During this period, the main thrust was to bring about institutional changes and to create agricultural infrastructure. An attempt was made through the community development programme to acquaint a large number of owner cultivators with better technological practices including those associated with irrigation. During this period, agricultural output grew at an average compound rate of 3.13 per cent per annum. Area increases constituted the main source of growth and accounted for nearly 52 per cent (and yield 42 per cent) of the total growth of output.

In the second period, which extended from 1967-8 to 1990-1, the main emphasis was on increasing yield through the use of modern inputs and improved methods of production in selected parts of the country. The role of technology as a major input in agriculture was given explicit recognition, and yield growth rate became the dominant source of agricultural output. Thus, during this period, the contribution of productivity increase to total output growth rate of 2.81 per cent per annum was above 90 per cent while that of area increase declined to less than 10 per cent.

An important feature in the pattern of agricultural growth which needs underlining is that because of differences in resource endowments and in the intensity of use of new seed-fertilizer technology, there exist marked variations in the level and growth rates of agricultural output recorded by various States of the country.

An analysis of Statewise data shows that in the periods 1962-5 and 1986-9, the highest rate of growth of agricultural output was achieved by the north-western States of Punjab, Haryana, Jammu and Kashmir and Uttar Pradesh. Next to the north-western States, the southern States of Andhra Pradesh and Karnataka had a consistently good performance. On the other hand, both Kerala and Tamil Nadu did not fare very well.

The pattern of agricultural growth is characterized by wide year-to-year fluctuations in various States located in the central region. For example, Gujarat which had a consistently good growth record up to 1980 faced severe drought later which resulted in a big dip in its growth rate during the eighties. Taking the entire periods 1962-5 and 1986-9, Rajasthan recorded a growth rate of 2.59 per cent and Madhya Pradesh and Maharashtra each exceeding 1.50 per cent per annum, but in all cases there were large year-to-year fluctuations in agricultural output.

Finally, till the beginning of the eighties, it was the eastern States of Assam, Bihar, Orissa and West Bengal which were the vulnerable areas of Indian agriculture. However, thanks to a spurt in rice production during the eighties, the growth rates accelerated in a perceptible manner. In particular, Bihar and West Bengal which were earlier the two slowest-growing, if not stagnating, States, have registered notable acceleration in their growth rates since the period 1980-3.

The wide differences in the growth rates of agriculture have had a differential impact on the living conditions of the peasantry in various parts of India. Thus, in the fast growing north-western region comprising Punjab, Haryana, Uttar Pradesh and Jammu and Kashmir, not only has output increased but labour productivity has also recorded a rapid growth. It can be safely presumed that the income levels of agricultural workers have tended to increase. In the mid-growing States, the output growth has hardly kept pace with the growth of population and therefore the per capita productivity and income have remained more or less constant. In large parts of central and eastern India and a few States in the south, however, the male workforce in agriculture has grown at a faster rate than the growth of output, thus leading to a decline in productivity per worker. It is in these areas that a process of involution has taken place and prima facie there are reasons to believe that the extent of poverty and destitution has tended to increase significantly.[5]

Two different political processes have emerged in the developed and underdeveloped parts of India. In the richer agricultural regions, increased incomes have helped to increase the extent of the market and have thus laid the basis for industrialization. However, because of certain major constraints, the process of industrialization and diversification has not been very rapid. On the political side, the power and influence of the richer farmers has increased tremendously. Simultaneously, the bargaining power of the landless labourers has also improved as a consequence of greater demand for labour. The emergence of the rural proletariat as an organized force is bound

5 Bhalla, G.S. and Tyagi, D.S., 'Spatial Patterns of Agricultural Development in India', *Economic and Political Weekly*, Bombay, June 1989

to have a far-reaching effect on the institutional framework in the countryside.

Furthermore, in these prosperous regions, the rich farmers having acquired economic benefits are gradually extending their political power. It is true that, to begin with, these farmers pioneered the adoption of new technology by investing in a big way in agriculture. This investment was partly financed out of savings and partly from borrowings from government agencies. So, to a considerable extent, the big increase in agricultural output can be attributed to their initiative and enterprise. But, having once acquired the essential assets, they are no longer investing their surplus in the same magnitude as earlier. Nor are they interested in investing in non-agricultural industrial activities about which they have little knowledge. Consequently, the surplus is now being wasted on conspicuous consumption, luxurious houses, extravagant marriages and social ceremonies etc.

At the same time, having become politically powerful they resist any attempt at mobilization of surplus by the State through increase in the supply prices of highly subsidized power, fertilizers and irrigation. They are also able to pressurize the government successfully in giving them higher procurement prices for major crops. To begin with, these subsidies helped to promote the adoption of new technology. These were specially beneficial for small and marginal farmers who would not have derived proportional advantage from increase in output prices. The subsidies also exerted downward pressure on the prices of foodgrains and thereby benefited the poor consumers in general. But over the years, the quantum of subsidies increased tremendously. The subsidies to the agricultural sector on fertilizers, electricity, irrigation and credit were estimated at Rs 11,793 crore for 1986-7 and have been steeply increasing every year.[6] Such large subsidies have drained the resources of the Exchequer and have thereby adversely affected public sector investment in rural infrastructure. Thus, the emerging kulak power is gradually becoming a barrier to the process of further capital accumulation in the agriculturally advanced regions of India.[7]

6 Gulati, Ashok 'Input Subsidies in Indian Agriculture — A Statewise Analysis'. *Economic and Polical Weekly*, June 24 1989.

7 Nadkarni, M.V. *Farmer's Movements in India*, Delhi, Allied Publishers, 1987.

On the other hand, in the rest of India, where agriculture is still underdeveloped, the condition of the bulk of small and marginal farmers and agricultural labourers continues to be pitiable and lack of dynamism has tended to keep the total market limited because of stagnating agriculture. Hence, it has not been possible to lay down the very basis for the process of industrialization. These conflicting and contradictory processes constitute the political economy of Indian agriculture at its present stage of development.

RURAL TENSIONS AND THE PEASANT MOVEMENT

During the colonial period, the main strategy of the peasant leadership was to unite all sections of working peasantry and landless labour against the rural feudal interests, the moneylenders, and the colonial regime. For the peasant movement, the most important contradiction existed between the peasantry on the one hand and the imperialist regime supported by the landlords and moneylenders on the other. The contradictions within the peasantry, and between cultivators and landless labour, were played down as these had not yet emerged in an accentuated form. The non-monetization of the agricultural economy, its relatively backward technology of production, and its imperceptibly slow progress, meant that the objective basis for differentiation, between cultivators and agricultural labourers, and between the small and big peasants, had not yet fully developed. The mode of production remained primarily pre-capitalistic, traditional and semi-feudal. The landless labourers, like the artisans, were governed by the traditional *jajmani* system and the village proletariat had not developed the characteristics of free labour much less those of a class in itself and for itself. The main objective of the peasant movement was to wage an anti-imperialist and anti-feudal struggle while carrying the entire peasantry (including landless labourers) with it in this task. Consequently, even though the slogan of class struggle was advanced for propaganda and mobilization, except for anti-feudal struggles, hardly any issues were taken up that sharpened the class struggle within the peasantry. In the context of pre-capitalistic agriculture and undifferentiated peasantry, such slogans would have been premature, anyway.

This colonial context has undergone a basic change since Independence. Firstly, the independence struggle has ceased to exist as a focal point for peasant mobilization. Secondly, with the partial success of land reforms during the fifties and early seventies, and the abolition of intermediaries, anti-feudal struggles have also lost their sharp edge in most parts of India. The main exceptions are areas like Bihar, Orissa, Rajasthan and Madhya Pradesh, where even intermediaries have not been abolished and where semi-feudal bondage relationships, though weakened, are still quite powerful. Further, although lip service is paid to implementation of land ceilings by almost all progressive parties including the ruling party, in actual practice there is hardly any concerted peasant struggle that has highlighted this aspect. Thirdly, agrarian relations are undergoing significant changes as a consequence of the very rapid growth of agriculture in some States. One important result is the emergence of the rural proletariat as a class, and signs of differentiation of the peasantry. This is specially so since the advent of the new technology in agriculture in the mid-sixties. Another equally important result is the emergence of kulak power and the consolidation of owner-cultivators under the leadership of rich farmers. Fourthly, some important changes are also taking place in the backward agricultural states, where some sections of the peasantry have recently acquired relatively more economic power and are now clamouring for more political power, leading thereby to both caste and class conflicts.

The most significant structural changes are taking place in the States of north-western India like Punjab, Haryana, western UP, eastern Rajasthan, the southern coastal strip of Andhra Pradesh, Tamil Nadu, and Kerala where basic changes have taken place in the technology of production and also in the production relations in agriculture. Firstly, mechanization in the form of tubewells, tractors and threshers has taken place on a large scale. Secondly, an overwhelming proportion of cultivators are engaged in commodity production and they take most of their produce to the market. Thirdly, because of the intensity of operations and the need for their timeliness, the demand for labour has increased tremendously and even the small cultivators have to hire labour at peak times. Fourthly, an increasing demand for labour and rise in real wages has also led to increasing mechanization. Fifthly, because of its profitability the new technology has spread quite widely both among various strata of peasantry, as also to various regions that are endowed with irrigation. The spread has become cumulative because of large investments undertaken in irrigation, land reclamation and other infrastructure. The rapid capital accumulation in agriculture was financed by both surplus generated on farms and by funds provided by state agencies at subsidized rates. Sixthly, rapid growth in agriculture led to an even higher growth in the secondary and tertiary sectors because of input, output and market linkages. Although for quite some time the diversification of the economy was hardly noticeable, of late there are visible signs of accelerated diversification of rural economy in the agriculturally developed regions of the country. Finally, because of increased productivity, the output and income levels of cultivators and of landless labour have recorded significant increases. Further, the gains of agricultural development have been shared by all categories of cultivators, albeit more or less in proportion to the distribution of land ownership. It is, however, the surplus-generating large and middle farmers who have been the greatest beneficiaries. The emergence of a lobby of powerful rich farmers has been the cause of several socio-political problems. On the other hand, another equally important development has been an increase in real wages of landless labour. The increasing demand for labour has enabled landless labour to break out of the shackles of the age-old *jajmani* system and emerge as free labour.[8]

The major cause of tensions in the agriculturally developed areas is the increasing assertion of landless labour of their rights and a demand for higher wages. The increased demand for labour has improved their bargaining power. This has also given them more freedom as they are no longer tied to a few cultivators. The new forces of 'bonds' through interest-free loans are more an expression of anxiety on the part of cultivators in need of assurance of help in the peak season than a relation of servitude[9] The result is that the landless are no longer mute — they are acquiring a new dignity. This is naturally very offensive to the sensibilities of land owners — who

8 This has been characterized as the capitalist mode of production by some scholars. We have deliberately avoided the controversial debate on modes of production in Indian Agriculture. For a comprehensive review of this debate see, Thorner, Alice, 'Semi-Feudalism or Capitalism? Contemporary Debate on Classes and Modes of Production in India' *Economic and Political Weekly:* December 11 and December 18, 1982.

9 Bhalla, Sheila, M E, 'New Relations of Production in Haryana Agriculture'. *Economic and Political Weekly*, 1976, 11 (13): p. A23

are used to treating them like dumb cattle. Consequently, the conflict between wage labour and cultivators has tended to increase and the major contradiction that has emerged is the one between agricultural labourers as employees and cultivators as employers.

On the other hand, land has ceased to be a major contradiction and there is hardly any movement, either of tenants against landlords, or peasants against moneylenders. Nor does one find any contradiction between the small and big peasantry despite a big difference in their economic status. The small farmers have also benefited as they are also able to share the benefits of the new technology. However, the capacity of small farmers to compete with big capitalist farmers is getting limited because of their very small resource base. Nevertheless, despite limited economic standing, they are a decisive force in determining election results because of their numerical strength. With the polity becoming increasingly populist, political parties vie with each other in pleasing the peasantry through input subsidies, loan waivers and so on. The small farmers are also able to obtain a share of subsidized inputs and other concessions. Normally, the big fish would have eaten the small fry. But the State has intervened effectively to sustain the small farmers. This has counteracted any possibility of contradictions developing between small and large peasants. On the other hand, in many areas a large proportion of small and marginal farmers are forced to hire out their family labour, as their tiny holdings are incapable of giving them sufficient income for subsistence. This has created conditions for these small farmers to form a united front with the landless labourers and has laid the objective basis for the differentiation of peasantry in future. But at present, both because of status, and because of caste considerations, small and marginal farmers have become willing, if not enthusiastic allies of large farmers. They are hesitant, if not outrightly opposed, to closely leaguing with landless agricultural workers who overwhelmingly belong to the scheduled castes. One of the biggest challenges facing the peasant movement is to bring about the unity between landless labour and the small and middle peasantry in areas of the 'green revolution'.

In some parts of the country, particularly in Punjab, rising religions fundamentalism has added a new dimension to the rural tensions. It is not within the scope of this paper to dwell at length on this issue. However, apart from various political and administrative causes, one vital reason for the alienation of a large number of village youth has been the inability of the growth process in the economy to provide decent productive job opportunities to an increasing number of semi-or half-educated (matriculates or plus two) rural youth. Rapid agricultural growth in these parts has not been accompanied by a rapid growth of a labour-intensive process of industrialization.

The character of contradictions and tensions is quite different in the less developed parts of India like Bihar and Orissa. In these parts the mode of production still remains semi-feudal and land distribution and abolition of intermediaries continues to be a major issue. However, even in these States there are certain sections of the peasantry who have been able to benefit from the new technology. These are the small and middle cultivating owners who, in general, belong to the lower and middle castes. Having become economically stronger, they are now clamouring for political power which at the moment vests in the semi-feudal upper castes. In the process of this struggle, caste and class tensions are getting accentuated. The emerging lower and middle castes are in no way less tyrannical towards landless labour than the big absentee upper-caste landlords. Hence the result is a two-way contradiction. On the one hand, there is strife between the landless and the nouveau riche and, on the other hand, there is a contradiction between the middle peasantry and the feudal interests. The caste factor has further complicated the situation. In many cases, because of convergence of caste and class, class conflicts take the shape of caste conflicts and vice versa. Very often caste, rather than class is used as a weapon for mobilization and for attacks and retaliation. Some radical parties, although wedded to class struggle because of sheer opportunism, do not consider it inappropriate to use caste-mobilization as a vehicle for election or political power.

It is important to remember that in some regions, particularly in Bihar, a very powerful movement of agricultural labourers, tenants and small peasants has emerged recently. Simultaneously, the aggressiveness of absentee landlords (who are gradually losing their economic and political power) along with that of the new rich of the village, has increased manifold. It must be said that the power of organized peasantry to fight back has also increased sufficiently. All this has resulted in heightened tensions leading to serious conflicts, including large-scale murders and counter-murders vitiating the entire atmosphere. These feudal contradictions are in sharp contrast to the situation in the established 'green revolution' region, where political power has already passed on to the rich peasants.

CONCLUSION

To sum up, Indian agriculture seems to be in a state of transition where the old semi-feudal land relations are, by and large, being replaced by self-cultivating peasant proprietorship. That the age-old stagnation in agriculture was replaced by truly dynamic agriculture in some regions after Independence is clear from the fact that the post-Independence growth rate of agricultural output was more than three times that achieved during the first three quarters of the twentieth century. Land reforms, creation of irrigation infrastructure and adoption of modern technology are mainly responsible for this breakthrough. The destruction of semi-feudal relationships created the objective conditions for the development of agriculture through the creation of a huge infrastructure in irrigation and through the availability of modern technological inputs. This has enabled the peasantry to reap the benefits of increased productivity and in some regions to grow on capitalist lines. In other regions, the semi-feudal relations, though considerably weakened, still constitute powerful barriers to widespread initiation of the growth process. In still some other parts, lack of irrigation infrastructure has hampered any agricultural breakthrough.

In the prime 'green revolution' areas, where the entrepreneurial farmers have adopted modern scientific techniques with a view to maximizing their profits, the growth has by and large been on capitalistic lines. The State has helped the growth process in a big way through the creation of basic irrigation infrastructure and provision of modern inputs at highly subsidized rates.

The capitalistic growth process itself has generated several contradictions, the most important being that between the ever increasing bulk of landless labourers and large cultivators. Furthermore, co-existence of very large, small and marginal farmers has created a situation of great instability. All these factors have contributed to the accentuation of rural tensions in many parts of India.

It would be naive to argue that development of capitalism in Indian agriculture is a retrograde step or that capitalist development in agriculture has exhausted its potentialities and already reached a state of crisis. However, it needs to be underlined that the capitalistic process of growth has generated quite serious contradictions. These have been exacerbated due to the fact that a solution to the entire problem of providing productive employment to a fast growing agricultural population is being sought in agriculture itself. Agricultural growth has not been very rapid everywhere. Even in areas where agriculture has proliferated, this growth has not been accompanied by a rapid labour-intensive process of industrialization. The result is a rapidly swelling pool of educated unemployed in the rural areas and a consequent rise of lawlessness, conflicts and religious fundamentalism. The lack of rapid diversification, accompanied by increasing population pressure on land, is resulting in accentuated rural tensions.

It is becoming clear that these contradictions can only be resolved through a far-reaching and radical reorganization of land relations and the shift of a large section of the population from agriculture to productive non-agricultural occupations. This would obviously depend on the ability of the democratic movement to bring about the required restructuring of the Indian economy and polity.

Reconstituting Indian Society

SATISH SABERWAL

As we approach the twenty-first century it is useful to seek the meaning of what has been happening and to ask where we should go from here. This essay braids four arguments together:

1) When in 1947 we inherited the Indian State, we had virtually no experience of constructing or maintaining a large State whose legitimacy would rest on the consent of the people, actively expressed. Consequently, we had very little by way of clear principles, or judgments, derived from our own history concerning the limits which must not be crossed. We did not establish a monitoring system which would know the heavy, cumulative costs of allowing a long rope to sundry adventures — e.g. Sanjay Gandhi, Bhindranwale, and Devi Lal. Our political managers did not know that such adventurers must be stopped in their tracks, let alone knowing how to move assuredly to do so.

(2) All societies have ordering devices. The necessity for such devices increases as a society becomes more complex, and the scale of its activities and the size of its organizations increase. These devices consist partly of general rules or ideologies, and partly of the apparatus of coercion. The less effective the available general rules or laws of ideology in securing orderly functioning are, the greater the play of coercion in the society.

(3) Orderliness in late colonial India resulted partly from established routines of colonial society, and partly from the caste order. Coercion, partly as built into the caste hierarchy, had been familiar in the lives and experiences of most people, especially in rural areas.

The effectiveness of these ordering devices has been declining rather steeply in recent decades.

(4) No tears need be shed over the erosion of either the caste or the colonial order. Many of us have gone much farther, however in attacking the processes of orderly governance themselves, often lauding every protest and every uprising as expressions of 'people's power'. The moral initiative in public opinion was preempted by voices that appeared to come from committed anarchists.

Only a thin, Westernized stratum has understood Western ideas of a society governed by general laws — the ubiquitous courts and lawyers notwithstanding. With older constraints eroding, and without access to alternative ideas and orderly skills for achieving one's objectives, the use of coercion and force — by sundry 'liberation' forces and the like, and by the Government — has been increasing correspondingly. How did this happen?

Let us recognize immediately that the difficulties in India are of a piece with difficulties elsewhere, and have been illustrated powerfully in the happenings in the former Soviet Union. We are all caught in a vast historic process. For good or for evil, it is Western Europe that worked out early the institutions and the technologies which today set the terms for everyone else. Vastly enlarged scales of possible social relationships, of production, of learning, and of much else, are central to this evolving universe. Alternative visions, whether the Marxists' or the Third World nationalists', have been losing steam: and as their protagonists try headlong to try to catch up with the West, they find themselves stumbling all too often. This is the general milieu within which one has to locate the particular Indian case.

The colonial State, in India as elsewhere, was created to European designs and was monitored by European skills. In the colonial enterprise in India, some Indians were fitted in as cogs in the colonial machine. Outside that machine, their principal political experience was in the national movement: as *opposition* politicians. Men like Gandhi, Nehru, and Patel shone in that role because they could speak the language of the British: not only English, literally, but, more important, the language of law, that is, the language of law as it developed in Europe since the eleventh century, with a long historical commitment to justice, and a public life informed by a rather humane vision.

India's confident transition into the democratic process after 1947 — contrasted with happenings in some other countries — flattered us with the illusion that, as a people, we were to the manner born. In fact we were not able to see through our own illusion. We underrated the virtues of the colonial State, taking its notions of orderliness to be pointlessly finicky; and at times we have given the impression of being in a hurry to banish these notions. The social characteristics of political leadership underwent a sea change meanwhile, as the very credentials needed for entering the ring changed: where the organizing of a large movement, jail-going, and defying the British had been crucial earlier, the new order looked rather for the ability to get the votes which were cast in the polling booth and to raise the funds which were needed for contesting elections.

The social sciences contributed to the nurturing of our illusions. The mobilizing of caste groups could be an element in the making of electoral majorities and, therefore, it seemed almost that the caste order was good for democracy. The American political sociologists, Lloyd and Susanne Rudolph, launched the famous phrase: 'the modernity of tradition'. There is a general lesson here. All of us — academics, journalists, politicians — got hooked on short-term patterns and acted as if these could be projected forward into eternity. Attempts elsewhere at weighing these short-term patterns against the longer term historical experience of other societies, such as Gunnar Myrdal's, tended to be dismissed as Euro-centric, neo-colonial, or whatever.

The shift in the operative criteria of political selection brought numerous hitherto dormant social categories erupting into the public arena. They arrived with only limited Western education, and with even less understanding of the nature of orderly processes in bureaucracies, the judiciary, universities, and the like : and of the strong, widespread sense of limits, so crucial to their optimal functioning. The grasp of these matters in these ascendant groups was even weaker than that of groups whom they were displacing. The whole arrangement — the institutions of Western inspiration, and the social stratum which had learned to operate them with some skill — came under growing pressure; and *there was no vision for viable alternative arrangements to replace it.* The field was wide open for all manner of adventurers to set up their own games, some more bizarre than others.

Yet, as I said, any complex society must have adequate ordering devices, or it will go under. If the Western ideas of orderliness did not make much sense in our milieu, what could we draw from within our own traditions? The caste order had placed the various castes in a pecking order in each locality, with each caste responsible for order within its boundaries through the working of caste panchayats and the like. Kingship had been another major source for orderliness, the king intervening when local groups could not sort things out on their own; and a certain amount of orderly functioning had been centred on sects, temples, and the like.

It is the caste order which survived the attrition of the encounter with the West better than other mechanisms, and it came to be supplemented by Western institutions implanted under colonial auspices. The caste order, however, came under powerful assaults; from above as well as from below it was an unmitigated evil.

We failed to notice that, however inequitously, it had provided also a large part of whatever social cement there was holding this acutely segmented society together. We have tended to go by the unstated assumption that our people will spontaneously generate whatever ideas and mechanisms they need for living happily together. We have not bothered to test how much of this assumption was merely wishful thinking.

B. R. Ambedkar, whose centenary was celebrated during the year 1991, was engaged in these assaults both from above and from below. From above, his influence, as Minister of Law when the Constitution of free India was being drafted, militated towards an ideological shift against the caste order and towards a regime of reservations for hitherto deprived social categories in various governmentally-controlled institutions and fields. This ensured accelerated mobility for at least some among the lower castes. From below, his urgings, as a leader who made the lowest castes his special constituency, called for determined efforts to flatten the caste system, leading to (may I say, a welcome) assertiveness on the part of the lower castes.

This assertiveness has been only one element, however, in a widespread spirit of insurgency which has found other sources of inspiration too: in the Gandhian legacy of civil disobedience, and in the Marxian challenge to authority — which, in India, has tended often to equate all authority with authoritarianism. The recent mushrooming of numerous other bases of belligerence, expressed in a great deal of retail violence by terrorists and others, is known well enough.

In this milieu of growing belligerence, the contemporary Indian State is caught in an extraordinary historic bind. The tradition of governance from pre-colonial times did recognize the importance of the king or the ruler for maintaining order. Bhiku Parekh has shown, however, that the Indian tradition did not generate anything that might be called political philosophy (The *Arthashastra* is not political philosophy, whatever else it might be).[1] The principal ordering device it provided the ruler with was the use of *danda,* deterrent and even excessive punishment. The tradition did not conceptualize the nature of the elements and the purposes which go into struggles for power. Put otherwise, it has been central to Indian political traditions to secure one's dominance over others by means of force and, furthermore, to use condign punishment for those who would defy the ruler. The idea of *danda* is in disrepute however, smacking of repression, authoritarianism, and the like — and therefore loathsome. The Indian State has indeed employed a vast array of armed force over the decades to handle the troubles in areas of high insurgency, but the effort is seen to be unacceptably oppressive by its victims, by the civil rights groups within India, and by an articulate international audience.

The State's legacy on this account includes legislation, courts of law, bureaucracy, and so forth. Underlying them is a logic of impersonal functioning, in terms of general principles; and this logic, when practised, has given these institutions their great historic power. For us, however, this logic is an alien, colonial conception. It has not arisen out of our historical experience,[2] and it has

1 Bhikhu Parekh, 'Some Reflections on the Hindu Tradition of Political Thought', in Thomas Pantham and Kenneth L. Deutsch, eds., *Political Thought in Modern India,* Sage,1 1986. See also Ranajit Guha. 'Dominance without Hegemony and its Historiography, *Subaltern Studies VI: Writings on South Asian History and Society,* Oxford University Press 1989, pp.210-309.

2 Indian society has been notable for the strength of its intra-familial and intra-caste bonds and for its difficulties in constituting public authority on a durable base. The enormous odds against sustaining impersonal normative orders under these conditions may not be underrated; see Leigh Minturn and John Hitchcock, *The Rajputs of Khalapur, India,* New York: Wiley, 1966, p.51.

commandedamonguslessandlessofeitherunderstandingorrespect with each passing decade.

Yet we give little thought, let alone effort, to what ought to be our collective responsibility for creating and maintaining conditions for reasonable order around us. Most of us are willing to leave it all to the Government, or whoever else stands for authority. In any case, we have reached a point where orderliness is at a large discount, and multilateral recourse to force wide-spread.

Central to a democratic, constitutional political regime is the institution of elections: the key mechanism for constituting legitimate authority. What has been happening to the institution of elections in India? An election is a device for picking between competitors. The assumption is that the sum of individual opinions of those who are to be governed amounts to the voice of the people; and the mandate from that voice has a sanctity, especially when there is no alternative basis for choosing legitimate rulers.

We must remember, however, that this device comes to us from the West. It has no deep roots, and therefore no great sanctity, in our ancient traditions. It has had to contend increasingly with ideas which go much deeper in our tradition, viz. the use of force, to establish one's dominance over others. In pre-colonial India, a group would have used its prowess at arms to establish its dominance in a village or a cluster of villages: this was the dominant caste, which controlled the land, and would command a large share in the produce of the land. On successively larger scales, the Raja's kingdom and the Imperial centre in turn were established through the force of arms, through conquest. *There was no other way.* The preferences of those who would be ruled counted for little.

These latter processes were, of course, in abeyance during the colonial period: the British maintained an effective monopoly over force. After 1857, on the rare occasions that this monopoly was challenged, they would reassert it ruthlessly. Gandhi's genius lay in outwitting them by eschewing recourse to force,effectively denying them thereby the option of military repression. The State's monopoly over force has been central to the making of Europe since the middle ages. Within that monopoly, they devised the mechanisms for political contention which we know as 'democratic political structures'.

These mechanisms arose in a particular cultural milieu however, and we should glance at that milieu briefly. The idea of the individual is important in it. It has many aspects: among them is that each person should enter the booth and express one's opinion in privacy by casting the vote. Secondly, the process of elections is governed by general rules and laws applied to everyone impersonally; and these are, of course, part of a whole universe of rules and laws intended for general, impersonal application.

Indeed the widespread use of impersonal general laws, codified systematically, has been the basis for justice in Western States. This assurance of justice has been central to the Western societies' ability to achieve a wider sense of community. Parenthetically, may I note also the historic importance of the Church in fostering this sense of community: it promulgated a shared creed, myths, rituals, and symbols; it supported marriages even with recent immigrants in a locality; and at critical moments it provided wider political and other public leadership. Altogether, the Church has contributed greatly to a tradition of open-ended social relatedness in the West.

Finally, the tradition of justice gained strength in the West because adequate numbers of persons were willing to work at administering justice for their fellowmen. The willingness to take on these responsibilities has been part of a wider ethic of public responsibility. One enters these duties not to line one's own pockets but to act for the common weal, for advancing a wider vision. Entering the Church was an early option for dedicating one's life to a larger cause. Something of the same spirit prevailed in teaching, in the civil services, in health-care, and in all manner of voluntary service. Marx, those who organized the Labour Party in England, and others, had this larger vision of a better world which would not pull the working classes down. Marx's vision soured ultimately; but meanwhile it had released enormous energies which literally moved mountains.

The mechanism of elections, for deciding who would rule, emerged as part of a particular, Western cultural matrix. The historic premises for Indian society had been very different. That rulers should be elected through the uncoerced expression of everyone's opinions was an extraordinary proposition whose general acceptance could not be taken for granted as part of a 'natural' order of things. Yet our options were few: conquests, or descent from conquerors, have been an alternative source of legitimacy; and, in recent times, military coups. We could not plan for either course. We needed, rather, to attend to the groundwork of ideas and relationships which make up our society so that, together, we might learn to live in more orderly ways.

The process of determining who should rule over us legitimately is important; and it is equally important that this process be orderly. Over the past decades people found that they could play tricks during elections, and their tricks became increasingly brazen. If you won an election, even if you cheated while doing so, people accepted your rule anyway. So people began to cheat more and more, and to use force. There are always thresholds in human affairs, however; and an excess of cheats and bullies means that the system cannot be trusted any more. Persistent cheating and despotism is liable to lead to fights and, if on a national scale, to civil wars.

Where, then, do we go from here? In these privatizing times we cannot get away with a pronouncement of the form, 'The Government must immediately...'. The task before us is nothing short of having to reconstitute our society, and that simply is not the politicians' or the economists' cup of tea. Efforts for this order have historically been sparked by prophets — Mohammed, Buddha, Nanak, etc. Our times are perhaps too democratic for genuine prophets to emerge. With all due respect to the likes of Swami Agnivesh, one must recognize the sharp disjunction between our Westernized intelligentsia and the institutions of modern learning on one side, and our men and women of religion on the other. Few of the latter have a clue to the scale of our society, or to the dynamics of the world we inhabit.

In such times, the obligation to take initiatives and to accept responsibilities of extraordinary magnitudes falls on ordinary women and men. No one can outline the agenda for reconstituting a large and complex society like ours in a few brief paragraphs. I sketch a few areas here, illustratively, hoping to provoke better responses from those whose insight is keener.

One key objective for us, it seems to me, has to be *justice.* Justice in the law courts is important; but we need justice too in myriad other settings and relationships: for daughters-in-law, for child workers, for the victims of technological change.... It is vital that our children should grow up with a sense of fair play, rather than narrow morality which sometimes makes one's own, or one's family's interests supreme.

The *managing of conflicts* is a related area. A world free of conflicts is an utopian's dream: the rest of us may seek at least to resolve them. Too much conflict arises because we project our floating anxieties and misfortunes on to others' presumed ill-will unreasoningly. All of us need to sharpen our skills at judging social situations and the sources of our misfortunes, individual and collective, in order to avoid the deadly traps of unrealistic conflicts.

Associated with both justice and conflict-resolution is the notion of *order,* considered earlier. I wish only to add that a society's capacity for creative achievement depends substantially on its ability to accommodate large, apparently unordered, spaces e.g., people's private lives; within powerful publicly ordered spaces like traffic lights, telephones and the flow of rivers.

Impregnating everything else, however, is the question of mutual *relatedness.* The caste order did traditionally simplify this question by virtually specifying with whom one could associate with, and how intimately. But now that the caste order is almost non-existent it is all the more necessary that we learn ways for relating to numerous others in myriad situations — ways for creating cordial, just, non-combative, varied human bonds in orderly ways. It is these endlessly ramifying bonds which together constitute society; and we need to work out, and show others in word and deed, how to limit the tensions and the lethal waste in our relationships, and how to augment their challenge and meaning and significance for each other. My actions do influence what life is going to be like: for my neighbour and, on the rebound, for myself.

I have noted here a sampling of the kind of ideas that await our urgent attention. Ideas, by themselves, are merely hot air or cold ink. These begin to make social differences, however, when embodied in and propagated by institutions.

Historically, in India, the range of our institutions has stretched over the caste order (including particular occupational niches, such as merchants' trading arrangements), very modest schools, and others that we have noted earlier briefly. Today, of course, we have a vast panoply of formal institutions; but all too frequently these get hijacked for private gain. We have not generated a shared ideology over how to operate and use our institutions, how to ensure that those who run, and work in, an institution abide by its purposes seriously. That is to say, we have not applied ourselves to the task of determining what the essential combinations of pressures, internal and external,[3] are for securing such institutional integrity.

For an institutionally-differentiated society, maintaining the purposive integrity of its institutions is of prime importance, as also is the maintenance of monitors which evaluate these institutions' adequacy continually, and of overarching visions which furnish the society with the motive force and the sense of direction over the long term.

3 'Internal' and 'external' here can refer to two boundaries : one, between a person's motivational, internalized controls as against the external, normative ones: the other, between and institution's internal control processes as against its monitoring in the wider society.

I thank Seminar, whose September 1991 issue carried this essay in an earlier version, for permitting the present adaptation

The State, Personal Law, and Islam in India

BAHARUL ISLAM

Indian society presents a mosaic of many religions. The Hindus constitute the largest religious community in the country, but there are several important religious minorities: Muslims (11.4 per cent), Christians (2.4 per cent), Sikhs (2 per cent), Zoroastrians (.02 per cent) and Buddhists (.07 per cent). Each is governed by a personal law of its own. The policy of the British Government in India was to allow each religious or sectarian community to follow its own personal or sectarian laws, customs and usage. But in the administration of justice, a common criminal law and a common law of evidence obtained throughout the country. Even the personal laws of the Hindus, the majority, and the Muslims, the largest minority community, were not codified, not to mention those of the smaller communities. At the turn of the century, D. F. Mulla, an outstanding jurist, who was incidentally a Parsi, broke new ground by compiling Hindu and Muslim laws and by bringing out *The Principles of Hindu Law* and *The Principles of Mohammedan Law* which are followed in Indian courts as definitive works on the subjects, although they are not codified and put on the statute books.

The corpus of religious legislation also included the Christian Marriage Act 1872, the Indian Divorce Act 1889 and The Parsi Marriage and Divorce Act, 1889. In 1929, The Child Marriage Restraint Act was passed, which applied to all communities. In 1937 was enacted The Muslim Personal Law (Shariat) Application Act. It did not codify Muslim Personal Law but did away with the customs and usage followed by Indian Muslims and provided that the Shariat would be operative in all courts in matters like marriage, divorce, maintenance, guardianship, intestate succession, gifts, trusts and *wakfs*.

The passage of the Dissolution of Muslims Marriage Act in 1939 was a momentous event. Under the Shariat law Muslim women had no right to dissolve their marriage though the husbands enjoyed the absolute right to do so unilaterally and at will. The 1939 legislation gave Muslim wives the right to divorce their husbands on certain prescribed grounds through the intervention of the court.

The Indian Legislature which passed the above-mentioned Acts also put through legislation for the Hindu community. These Acts are:

1. The Hindu Disposition of Property Act, 1916.
2. The Hindu Inheritance (Removal of Disabilities) Act, 1928.
3. The Hindu Gains of Learning Act, 1930.
4. The Hindu Women's Right to Property Act, 1937.

Thus some positive steps were taken towards codification in certain areas of classical Hindu law.

Personal Criminal Law or Personal Law of Evidence of the religious communities was not followed in the country. On the procedural side, the Code of Criminal Procedure and the Indian Evidence Act which were modelled on the English system held sway.

It may be mentioned here that the tribals of some North-eastern States such as Nagaland, Meghalaya, Mizoram and Arunachal Pradesh have during the British regime and even after Independence been following their own customary laws. Even the National Codes of Criminal Procedure and Civil Procedure do not apply to courts in these areas, including the Guwahati High Court (which is the common High Court of all the seven States of north-east India). Only their spirit prevails, for the people and their social organizations in the region are very simple.

In this essay, I shall confine myself only to certain Family Law aspects of Hindu and Muslim Personal Laws. The changes that have or ought to have come about in these areas since Independence have been brought into focus.

HINDU PERSONAL LAW

Before the post-Independence changes in Hindu law were introduced, the position of Hindu women was unenviable. In the eyes of the law, they had no status. They had no right to obtain separation from their husbands. They had no right to property. In childhood, they belonged to their fathers; in young age, to their husbands; and in old age, to their sons. Independence brought about a change. Orthodox Hindu Personal Law was superseded by what is popularly known as the Hindu Code consisting of four Acts, viz:

1. The Hindu Marriage Act, 1955.
2. The Hindu Succession Act, 1956.
3. The Hindu Minority and Guardianship Act, 1956
4. The Hindu Adoptions and Maintenance Act, 1956.

The passing of these laws did not reflect the Hindu consensus. On the contrary, it was done in the teeth of strong opposition from Hindu orthodoxy in the name of religion. But the Indian Parliament cut through the resistance. The entire mass of Hindu law was codified and the Hindu Code saw the light of day, ushering in a new era in Hindu society especially for Hindu women who now enjoy equal rights with men in Personal Law as envisaged in the Constitution. So far as the mundane aspects of life are concerned, Buddhists, Jains, Sikhs as well as Hindu sects, all come within the ambit of the Hindu Code.

THE MUSLIM PERSONAL LAW

Let us now turn to Muslim Personal Law. Before the advent of Prophet Mohammed, women of Arabia enjoyed a position no better than that of Hindu women under classical Hindu law. Under the Quoranic law as interpreted so far by orthodox Muslim interpreters, two women are equated with one man in every respect including the right to property. In spite of India's constitutional provisions for equality of the sexes, Muslim Personal Law enjoins inequality between men and women. Because of the existence of Article 44 in the Constitution of India, there is a persistent demand among a section of Hindus for a uniform Civil Code. On the other hand, there is an equally persistent opposition to such a code emanating from Muslim orthodoxy on the ground that a uniform

Civil Code will mean interference with the *shariat*. The Government will not bring in legislation for the adoption of a uniform Civil Code unless the Muslim community demands it. It is highly unlikely that such a demand will come from that community in the foreseeable future.

Let us now examine some of the important aspects of Muslim Personal Law.

(1) Marriage

Polygamy

According to orthodox Muslims, a Muslim male has a right under the *shariat* to have four wives at the same time. In fact, however, the *Koran* by implication enjoins monogamy. The relevant verses of the sacred text have been misinterpreted. These verses are:

- Of women who seem good in your eyes, marry one; two, three or four; and if you fear, that you will not act *equitably, then only one:... This will make justice easier.* (*Sura Al Nisa*: iv:3)
- *And you will not have it at all in your power to treat your wives alike, even though you would feign to do so.* (emphasis added) (*Sama Sura*, verse 129)

Although verse 3 of *Sura Al Nisa apparently* allows a man to marry four times, verse 129 holds out a serious warning which cannot be ignored. It is like a proviso which, read with the italicized portions of verse 3, cannot be complied with. Equal or inequitable treatment depends not on the subjective belief of the husband *but* on the objective reaction of the wives, the recipients of the treatment, who naturally have different likes and dislikes, needs and temperaments. Ask any educated woman if she would like to be a co-wife. Or ask any co-wife in private and in confidence if she is happy being a co-wife. In both the cases the answer is bound to be negative.

What is more, polygamy militates against the dignity of women. Turkey banned it completely in 1951 and Tunisia in 1964. Syria, Iraq and Pakistan have restricted man's right to acquire many wives. In these countries, if a man wants to marry again while his wife is alive, he has to obtain special permission from the Court which is given only on a valid consideration. In Pakistan, the first wife can seek dissolution of marriage if her husband commits bigamy. But in India there is no restriction on polygamy for Muslims. Educated

Muslims believe in monogamy. But illiterate, ignorant and poor Muslim women do not refuse to be co-wives on account of economic compulsions. Food and clothing are more important to them than anything else. The State maintains benign neutrality. It has not introduced legislation to prohibit or restrict polygamy among Muslims, lest the party in power lose Muslims votes in the election on the allegation of 'interference with the *shariat*'.

Prohibition of Marriage on the Grounds of Relationship

Under the existing Muslim Personal Law, a Muslim male may marry his first cousin. The Hindu Marriage Act prohibits it with a proviso. Section 3(g) of the Hindu Marriage Act, *inter alia,* permits marriage of 'the parties not within the prohibited relationship, *unless* the custom or usage governing each of them permits of a marriage between the two'. A man and a woman are said to be within the degree of prohibited relationship, (a) if one is a lineal ascendant of the other; (b) if one is the wife or husband of a lineal ascendant or descendant of the other; (c) one is the wife of the brother or of the father's or mother's brother, or grandfather's or grandmother's brother; (d) if the two are brother or sister, uncle and niece, aunt and nephew, or children of brother and sister, or of two brothers or of two sisters.

Presumably the bar to marriage between persons within the prohibited relationships is based on the modern science of genetics and heredity. This is a healthy law. Exogamy is conducive to the mental, physical and intellectual growth of the progeny. Endogamy, on the other hand, is detrimental.

It is interesting that on this point Quranic law is similar to the Hindu Marriage Act. The Holy *Koran* forbids marriage with:

> (a) mother (b) daughter (c) sister (d) paternal aunt (e)maternal aunt (f) brother's daughter (g) sister's daughter (h)foster mother (i) foster sister (j) wife's mother (k) step-daughter (l) son's wife.

Scanning the above list, one cannot find prohibition of marriage by a Muslim male with his father's brother's or sister's daughter, and mother's brother's or sister's daughter. Nor is there any prohibition of marriage with:

(a) mother's mother, or her mother and so on
(b) father's mother, or her mother and so on

(c) son's daughter, or her daughter and so on
(d) daughter's daughter, or her daughter and so on
(e) father's sister

But later on the *shariat* put restrictions based on an interpretation of Quranic law, as (a) 'mother' includes all female ascendants both on the father's and mother's side and (b) 'daughter' includes daughter of son or daughter, how-low-so-ever.

This was certainly a more scientific, healthy and welcome interpretation. Similarly, marriage between first cousins is ruled out if the following interpretation is accepted: 'sister' includes father's brother's or sister's daughter or mother's brother's or sister's daughter'. This, it is submitted, will be a more scientific and healthier way of interpreting the Quranic verse.

Divorce –Talaq

Talaq means unilateral divorce of a wife by her husband. According to the Shariat, the husband can divorce his wife without any reason and at any time. This has been accepted as a correct and valid Islamic law even by the High Courts in India since the beginning of British rule. This law is incorporated in the textbooks on law and is taught at law colleges in the Indian subcontinent.

The law providing for *talaq* is wrong and contrary to the Quranic injunction on the subject — I would refer to a judgment of my own on *talaq* while I was a judge of Guwahati High Court which covers all the seven north-eastern States. It was delivered on 31 March 1978 in the case of Jiauddin Ahmed *vs* Anowara Begum. The case has since been reported (1981) in the *Guwahati Law Report* 358 and in *Islamic and Comparative Law Quarterly* vol.II no.1, March 1982. After quoting extensively from the Holy Koran and referring to the opinions of several Muslim authorities on the subject, I held:

> A perusal of the Quranic verses quoted above and the commentaries thereon by well-recognized scholars of great eminence like Mohammed Ali and Yusuf Ali and the pronouncements of great jurists like Ameer Ali and Fyzee, completely rules out the observation of Macnaghten that 'there is no occasion for any particular cause for divorce, and mere whim is sufficient', and the observation of Batchelor J. [I.L.R. 30 Bom.] that 'the whimsical and capricious divorce by the husband is a good law, though bad in theology'. These observations have been based on the concept that women were chattel belonging to men, which the Holy

> Koran does not brook. Costello J. in 59 Cal. 833 has not, with respect laid the correct law of *talaq*. In my view, the correct law of *talaq*, as ordained by the Holy Koran, is that *talaq* must be for a reasonable cause and be preceded by attempts at reconciliation between the husband and the wife by two arbiters – one from the wife's family and the other from the husband's. If the attempts fail, *talaq* may be affected. ... an attempt at reconciliation by two relations, one each from the parties, is an essential condition precedent to *talaq*.

This decision of mine has been approved later as the correct enunciation of the law of *talaq* by a Division Bench of the same Court in the case of Must. Rukia Khatun *vs* Abdul Khalique Laskar which was reported in 1981 in the *Guwahati Law Report* 375.

The *talaq* law as it prevails in India and some Muslim countries is inhuman, unjust, inequitable and contrary to the Quranic injunction and therefore must be banned. In some countries it has either been abolished or restricted.

It was abolished in Indonesia in 1947, in Tunisia in 1956, in Egypt as early as the 1920s, in Turkey in 1951, in Algeria in 1959, in Iraq in 1963, in Iran in 1967 and in Pakistan in 1961. In these countries divorce can be granted only by the Courts.

In spite of its secularism, or perhaps because of it, India has not taken a step forward to protect helpless married women from the whims of their husbands who exercise the right of *talaq* as prescribed in the Shariat. It is high time India took action on this issue.

Maintenance of a Divorced Wife

Section 125 of the Code of Criminal Procedure, 1973, provides for speedy action in obtaining maintenance from a husband, *inter alia,* by a divorced wife if the husband has sufficient means and the wife does not. Unlike in the preceding Code of Criminal Procedure, 1889 (S. 488), Cr. P.C. of 1973 passed by the Indian Parliament after Independence, made an improvement by adding an explanation, in Clause (b) which defines a wife as:

> ... Including a woman who has been divorced or has obtained divorce from her husband and has not remarried.' It applies to a divorced wife of any community.

In the famous case of Shah Bano, who was divorced by her husband when she was over sixty, the husband took the plea that since he had divorced her, he was not liable to pay maintenance under the Shariat, except for the period of *iddat.* The Magistrate rejected his plea and ordered him to pay maintenance under Section 125 Cr. P.C. The case was referred to the Supreme Court which ruled that 'wife' in the Cr.P.C. includes a 'Muslim wife' and therefore a 'divorced Muslim wife'. It upheld the lower court's orders.

Muslim orthodoxy was up in arms alleging that the Supreme Court had interfered with the Shariat. Pressure was brought to bear upon the Government to undo by legislation the verdict of the Supreme Court. Rajiv Gandhi's Government yielded to the pressure. It was the Muslim Women (Protection of Rights on Divorce) Bill, 1986 which appeased the fundamentalists. The controversial Bill was passed by Parliament. It provides, *inter alia,* that the husband is liable to pay maintenance to a 'divorced wife' only during the period of *iddat* (about 3 months). As a member of Parliament, I vehemently opposed the Bill though I belonged to the Congress(I). I pointed out that the phrase 'during the period of *iddat'* was not enjoined by the Koran; it was interpolated by Muslim law-givers who were all male. It was highly prejudicial to women. I also stated that the Bill was at variance with the provision of equality in the Constitution. But mine was a voice in the wilderness. It is respectfully submitted that the legislation is *mala fide* and *ultra vires* the Constitution. It is a colossal retrograde step taken by the Indian Parliament in the sphere of social legislation since Independence. It has segregated divorced Muslim women from divorced women of other countries and downgraded their status.

Interference with Muslim Personal Law

During British rule there had been a great deal of interference with Islamic law in India. The Islamic Criminal Law, substantive and procedural, and the Islamic Law of Evidence have been totally discarded. Neither the Koran nor the Shariat granted the right to woman to dissolve her marriage on any ground. The Dissolution of Muslim Marriage Act, 1939 has granted that right to Muslim women which can be exercised on as many as seven grounds. The Child Marriage Restraint Act, 1929, applies to Muslims as much as to other religious communities. According to the Act, marriage is not valid unless the bridegroom is at least eighteen and the bride is at least fifteen. Muslim Personal Law allows marriage if both parties have attained puberty which means both the boy and the girl are fifteen. The Indian Succession Act, 1925, embraces within its fold Indians of all religions. Likewise, the Special Marriage Act, 1954, permits marriage between persons of the same religion or of different religions, including Islam, or of no religion. Muslims of India never raised their voice in protest against these pieces of legislation. And the Acts have done the Muslim community no harm; on the contrary, these have been beneficial.

(2) Inheritance

According to Muslim Personal Law, a woman inherits half as much as her male counterpart. This is not based on the *shariat*. It is Quranic law.

It is necessary to give a rational interpretation of the relevant Quranic verse today in order to ensure that women get equal status as men without violating the provision of the Holy *Koran*, and, at the same time adhering to the prescription of the Indian Constitution.

To understand the Quranic verses on inheritance of property, it is necessary to take a look at the position of women before the advent of Islam. At that time women had no right of inheritance. The apparent meaning of the relevant Quranic verses is that a female will inherit half as much as her male counterpart. The five basic tenets of Islam have to be kept in view, viz, *kalima* (articles of faith like 'there is no God but God), *namaj*, *roja*, *zakat* and *haji*. The Quranic injunction on these five tenets is the same for both males and females. In other words, obligations of women on these tenets to Allah are the same as men's, not half. Both men and women have to offer prayers five times a day, both have to fast for a month during the month of Ramzan, affluent men and women have to pay *zakat* at the same rates, and financially sound and physically capable men and women have to go on pilgrimage to Mecca at least once in a lifetime bearing the same amount of hardship.

If a woman's obligation to Allah is the same as a man's, does it not follow that a woman should inherit an equal share of property as a man? Let me explain how it will not be an act of *gunah* (sin) but an act of *neki* (virtue) if a person says his or her prayers not five times a day butt more often. That applies to fasting during the Ramzan month of roj as well: *roja* can be performed for a longer period or more frequently to acquire greater virtue. *Zakat* can be doubled or quadrupled in amount, say from 2.5 per cent of property to 5 or 10 per cent. Similarly, if a woman inherits not half but as much as her male counterpart, it will be an act of *neki*, not of *gunah*. So, the rational and humanitarian interpretation of the Quranic verse in keeping with the spirit of modernity will be: 'Give the woman *not less than* half of her male counterpart's share. Give her an equal share.' This will also be in concurrence with the Equality Clause of the Constitution.

Unfortunately the State has taken no steps so far with regard to inheritance under Muslim Personal Law. It is high time the secular leaders of India addressed the problem in the manner that Jawaharlal Nehru did for Hindu women. Change will not be an interference with the *shariat* or Islam but it would be putting the Quranic verse on a rational basis.

If the Quranic laws are interpreted in the manner indicated above, it will be a great leap forward towards the adoption of a uniform Civil Code. Orthodox Hindus are demanding it on the strength of Article 44 of the Constitution. Orthodox Muslims are opposed to it on the grounds that it will be an interference with the shariat. I respectfully submit that the fear is baseless. Far from being an interference, it will mean an improvement.

A loophole in the Law of Inheritance

There is a loophole in the law of inheritance under Muslim Personal Law. It is that a child of a predeceased child of the owner of property cannot inherit it. This lacuna has been removed in Pakistan by the Ordinance of 1961 under which such a child inherits the predeceased parent's share. It is to be hoped that India will also legislate to remove this loophole.

(3) Different Schools of Personal Law

The Muslims regard the Koran as a Revealed Book containing the words of Allah. The Hindus regard the Vedas in the same way, the Christians the Bible, the Jews the Torah, and the Zoroastrians the Avesta. But these are not rigid books and that goes for the Koran as well. They are flexible and bear new interpretations to meet the exigencies of time and circumstances. These scriptures as originally written could not be expected to deal with problems that would arise several centuries later. That is why different scholars in different ages have put different interpretations on different provisions of the religious texts on which Personal Law is based: and consequently, there have been many schools of Hindu and Muslim law. At no stage of history has the Muslim legal system been uniform in any country of the world.

ISLAM IN INDIA

Indian Heritage

The position of Islam in India cannot be understood without a study of India's ancient heritage. This heritage teaches us tolerance, love, and a dedication to peace. These values derive from the message of the Vedas and the Upanishads.

At the beginning there was the Aryan and Hindu age which stretched up to the medieval times. As the medieval period began, Muslim invaders started pouring into India from the West and the North West bringing Islam to this country. But the Muslim invaders came to conquer and rule, not to spread Islam. At that time, Hindu society had no cohesion. Islam had a powerful impact on the fragmented structure, and numerous low-caste Hindus embraced Islam which offered an egalitarian society. Hinduism then was indeed a dharma (not a 'religion' as it is understood in English). Tolerance in Hindu society was unparalleled. Hindu dharma and culture was tolerant of newcomers as well as Hindu converts to Islam. A new Indian society was born.

At a later stage came the Christians and with them Christianity reached these shores. A large number of Indians accepted Christianity. Indian society, civilization, and culture assimilated the new force and experienced a renewal.

According to some scholars, the word 'Hindu' originally meant those people who lived on the eastern side of the Sindhu (Indus) river. It was the Chinese and Muslim travellers who gave them this name. So the term 'Hindu' encompassed Muslims, Christians and other communities living in that region. Religion therefore did not divide the Hindus. It was a compact and tolerant society.

It was the British rulers who divided Indian society into many segments in the name of religion and ethnicity. The purpose was to perpetuate their rule. They introduced separate electorates in the name of religion, sub-religion and ethnicity. The outcome was the partition of the Indian subcontinent into India and Pakistan on the basis of a 'two nation theory'.

The Muslim masses understood little or nothing about it, as the great majority of them were illiterate and politically exploited by their power-hungry leaders who later left for Pakistan to hold plum positions, while a large number of their co-religionists stayed in India.

The Muslim Fear

The basic fear of the Muslim community of India today is that of losing its identity. The Indian Constitution is expressly secular. The term 'secular' was not in the Preamble to the original Constitution of India. The Constitution (42nd Amendment) Act, 1976 incorporated it in the Preamble with effect from 31 March 1977 and today it is one of the cornerstones of the Indian polity.

Articles 25 and 26 guarantee religious freedom to all. Articles 29 and 30 grant protection of cultural and educational rights including the right of religious and linguistic minorities to establish and administer educational institutions of their choice. Articles 25, 26, 29 and 30 occur in Part III of the Constitution which provides for Fundamental Rights which are justiciable. Article 26 can be enforced at the High Court level and Article 32 at the level of the Supreme Court, the apex body. The prospect of obliteration of the Muslim identity, religious and cultural, is inconceivable. Barring miscreants, political parties and politicians in India are relatively free from the canker of communalism.

Every religion has two aspects: the spiritual, and the mundane. The spiritual relates to man's relations with his creator. The mundane deals with his worldly affairs which in many cases overlap with the spiritual. Those who believe in a Creator, call him God, Allah, Isvar, Jehovah or by any other name, and have an inner craving to know, please, attune to and reach out to the invisible Almighty. This craving is not thwarted by setbacks. Indian democracy is perceptive of this reality and the Constitution provides for freedom of religion. Even the Communist countries could not abolish faith in religion and its practice. In my view, one of the causes of the fall of the Communist regime in the Soviet Union was the attempt to suppress freedom, including religious freedom.

Islam as a faith, and as practised, will continue to live on in India. The Indian heritage confirms this belief. But because of communal disturbances in parts of the country, doubts have been expressed by some Muslims about the safety of the religious and cultural identity of Islam. Communalism and communal flare-ups are but passing phenomena. It is to be hoped that scientific education and economic development will put an end to them once and for all. The unholy alliance between greedy politicians and vested economic interests which help inflame communal passions will be busted.

Hindu Suspicion

But a lurking suspicion nags the Hindu mind. Hindus are assailed by doubts about the genuineness of Muslim loyalty to this country. Suspicion lingers about whether at a pinch, they will behave in the same manner as they did at the time of Partition. These misgivings are baseless, considering the following facts:

(a) Most of the pro-Partition leaders who were responsible for the division of the country are either dead or out of the country. A new generation of Muslims has been born.

(b) After Partition, Muslims have come to realize that the condition of the Muslim masses in Pakistan including the eastern wing, which is now Bangladesh, is worse than theirs. In India, they enjoy civil and political rights that are conferred on Hindus and other non-Muslims.

(c) Since Independence, India has been attacked by Pakistan twice, in 1965 and in 1971. Not a single Muslim betrayed India on either occasion. On the contrary, Havildar Abdul Hamid knocked out Pakistani tanks and gave his own life in the 1965 War. He was posthumously awarded the Param Vir Chakra by the President of India. Major M. A. R. Shaikh fought valiantly and died a martyr. He was posthumously awarded a Vir Chakra. Lt. Col. Salim Calch was also awarded posthumously, the Maha Vir Chakra.

Pakistan's machinations to arouse communal passions in Indian Muslims against their Hindu compatriots have failed. Indian Muslims have not responded to the battle-cry for Pan-Islam or Islamic brotherhood. They have seen Muslim nations tear into each other. Iraq and Iran fought for eight years; Iraq invaded Kuwait; most Muslim countries of West Asia, including Saudi Arabia, sided with the USA to fight Iraq, a Muslim country. Indian Muslims have seen through the myth that religion is a unifying force. It has been rightfully replaced by nationalism. They have also realized that Islam is an integral part of India's composite culture and nationhood. They are aware that distinguished Muslims assume high office here.Communal eruptions in India are not due to the religious differences between Hindus and Muslims. They are triggered by political and economic forces. But the political leadership of the country and the Government is doubtless non-communal and far from being anti-Muslim.